MEDICAL
INTELLIGENCE
UNIT

Molecular Mechanisms of Pathogenesis in Chagas Disease

John M. Kelly, B. Sc., Ph.D.
Department of Infectious and Tropical Diseases
London School of Hygiene and Tropical Medicine
London, England, U.K.

LANDES BIOSCIENCE / EUREKAH.COM
GEORGETOWN, TEXAS
U.S.A.

KLUWER ACADEMIC / PLENUM PUBLISHERS
NEW YORK, NEW YORK
U.S.A.

Molecular Mechanisms of Pathogenesis in Chagas Disease

Medical Intelligence Unit

Eurekah.com / Landes Bioscience
Kluwer Academic / Plenum Publishers

Printed in the U.S.A.

Kluwer Academic / Plenum Publishers, 233 Spring Street, New York, New York, U.S.A. 10013
http://www.wkap.nl/

Please address all inquiries to the Publishers:
Eurekah.com / Landes Bioscience, 810 South Church Street
Georgetown, Texas, U.S.A. 78626
Phone: 512/ 863 7762; FAX: 512/ 863 0081
www.Eurekah.com
www.landesbioscience.com

Molecular Mechanisms of Pathogenesis in Chagas Disease edited by John M. Kelly, Landes / Kluwer dual imprint / Landes series: Medical Intelligence Unit

ISBN: 0-306-47849-8

Library of Congress Cataloging-in-Publication Data

Molecular mechanisms of pathogenesis in Chagas disease / [edited by]
John M. Kelly.
p. ; cm.
Includes bibliographical references and index.
ISBN 0-306-47849-8 (alk. paper)
1. Chagas' disease--Pathogenesis--Molecular aspects.
[DNLM: 1. Chagas Disease--etiology. 2. Molecular Biology. 3. Trypanosoma cruzi--genetics. WC 705 M718 2003] I. Kelly, John M. (John Morrison), 1953-
RC124.4.M64 2003
616.9'363--dc21

2003012623

Dedication

To Anita, Sam and Ben. For keeping me smiling.

CONTENTS

EDITOR

John M. Kelly, B. Sc., Ph.D
Department of Infectious and Tropical Diseases
London School of Hygiene and Tropical Medicine
London, England, U.K.
John.Kelly@lshtm.ac.uk
Chapter 4

CONTRIBUTORS

Ivan D'Orso
Univ. Nacional de General San Martin
Inst de Investigaciones Biotecnológicas
San Martín, Pcia. De Buenos Aires
Argentina
Buenos Aires, Argentina
ivandor@iib.unsam.edu.ar
Chapter 3

Javier M. Di Noia
Univ. Nacional de General San Martin
Inst de Investigaciones Biotecnológicas
San Martín, Pcia. De Buenos Aires
Argentina
Buenos Aires, Argentina
jdinoia@iib.unsam.edu.ar
Chapter 3

Roberto Docampo
Laboratory of Molecular Parasitology
Department of Pathobiology
College of Veterinary Medicine
University of Illinois
at Urbana-Champaign
Urbana, Illinois, U.S.A.
rodoc@uiuc.edu
Chapter 5

Maria Carolina Q. Barbosa Elias
Departamento de Microbiologia
Imunologia e Parasitologia
Universidade Federal de São Paulo
São Paulo, S.P. Brazil
Chapter 2

Wenda Gao
Department of Medicine
Beth Israel Deaconess Medical Center
Harvard Medical School
Boston, Massachusetts, U.S.A.
wgao@caregroup.harvard.edu
Chapter 9

Michael Gaunt
Pathogen Molecular Biology
and Biochemistry Unit
Department of Infectious
and Tropical Diseases
London School of Hygiene
and Tropical Medicine
London, England, U.K.
Chapter 1

Marcella Faria
Departamento de Microbiologia
Imunologia e Parasitologia
Universidade Federal de São Paulo
São Paulo, S.P. Brazil
Chapter 2

Alberto Carlos C. Frasch
Univ. Nacional de General San Martin
Inst de Investigaciones Biotecnológicas
San Martín, Pcia. De Buenos Aires
Argentina
Buenos Aires, Argentina
cfrasch@iib.unsam.edu.ar
Chapter 3

Gislâine A. Martins
Department of Biochemistry
and Immunology
School of Medicine of Ribeirão Preto-USP
Sao Paulo, Brazil
Chapter 6

Michael A. Miles
Pathogen Molecular Biology
and Biochemistry Unit
Department of Infectious
and Tropical Diseases
London School of Hygiene
and Tropical Medicine
London, England, U.K.
michael.miles@lshtm.ac.uk
Chapter 1

Paola Minoprio
Institut Pasteur
Department of Immunology
Paris, France
pmm@pasteur.fr
Chapter 7

Silvia N.J. Moreno
Laboratory of Molecular Parasitology
Department of Pathobiology
College of Veterinary Medicine
University of Illinois
at Urbana-Champaign
Urbana, Illinois, U.S.A.
s-moreno@uiuc.edu
Chapter 5

Miercio A. Pereira
Tufts Medical School
Parasitology Research Center
Department of Pathology
Boston, Massachusetts, U.S.A.
maperrin@yahoo.com
Chapter 9

Rafael Marques Porto
Departamento de Microbiologia
Imunologia e Parasitologia
Universidade Federal de São Paulo
São Paulo, S.P. Brazil
Chapter 2

Julio Scharfstein
Universidade do Brasil
Instituto de Biofisica Carlos Chagas Filho
Cidade Universitaria
Rio de Janeiro, Brazil
scharf@biof.ufrj.br
Chapter 8

Sergio Schenkman
Departamento de Microbiologia
Imunologia e Parasitologia
Universidade Federal de São Paulo
São Paulo, S.P. Brazil
sergio@ecb.epm.br
Chapter 2

João S. Silva
School of Medicine of Ribeirao Preto-USP
Deptartment of Biochemistry
and Immunology
Sao Paulo, Brazil
jsdsilva@fmrp.usp.br
Chapter 6

Mauro M. Teixeira
Department of Biochemistry
and Immunology
Biological Sciences Institute
Federal University of Minas Gerais
Belo Horizonte, MG
Chapter 6

Shane R. Wilkinson
London School of Hygiene
and Tropical Medicine
Department of Infectious
and Tropical Diseases
London, England, U.K.
shane.wilkinson@LSHTM.AC.UK
Chapter 4

Matthew Yeo
Pathogen Molecular Biology
and Biochemistry Unit
Dept. of Infectious and Tropical Diseases
London School of Hygiene
and Tropical Medicine
London, England, U.K.
Chapter 1

PREFACE

It is almost 100 years since Carlos Chagas gave his name to the disease that results from infection with the insect-transmitted protozoan parasite *Trypanosoma cruzi*. This debilitating chronic condition continues to have a major impact on health in Latin America where it produces more than 4 times the combined burden of the other major parasitic infections. There is no current prospect of a vaccine against Chagas disease and chemotherapy is characterized by toxicity and limited efficacy. Recently, improved public health measures such as the "Southern Cone Initiative" have been successful in reducing parasite transmission in some countries. It is daunting to think though, even if all transmission can be blocked throughout the entire geographical range of the parasite, some individuals will still be suffering from this disease in 40 years time.

Chagas disease is a complex condition and the mechanisms of pathogenesis have long been an area of controversy. The progress of the human and *T. cruzi* genome projects, together with associated technological developments, now provide a new framework to facilitate dissection of the mechanisms of disease pathogenesis at the molecular level. It is timely therefore to review the current status of knowledge and to consider how these new findings will impact on drug and vaccine development. This book contains contributions from international specialists from across the spectrum of Chagas disease research, ranging from those with interests in the genetics and population biology of the parasite to those who focus on the fine-tuning of the human immune response. In combination these chapters present a picture of an active research community that is striving to understand and resolve a series of important and complex biomedical questions.

The *T. cruzi* species exhibits considerable diversity, and an association between parasite genotype and disease outcome has been widely considered a possibility. As outlined by Miles and colleagues, *T. cruzi* has now been subdivided into two major genetic lineages (I and II). There is some evidence that the more severe clinical outcomes result from infection with lineage II, but given the extent of variation within lineages, even this may be an oversimplification. In a landmark discovery these authors also report experimental proof of genetic exchange in *T. cruzi*, thereby resolving another longstanding debate. As discussed, this will have major implications for parasite population genetics and disease epidemiology, since it provides a possible mechanism for the generation and spread of novel phenotypes that could include enhanced virulence and drug resistance.

Gene expression in trypanosomes displays a number of unusual features that can impinge on disease outcome, the best characterized example being the antigenic variation phenomenon that facilitates immune evasion by African trypanosome. In *T. cruzi* it is becoming clear that gene regulatory mechanisms also have an important role in the interaction between parasite and host and that a greater understanding of the mechanisms involved will shed light on aspects of disease pathogenesis. Schenkman and colleagues now describe two distinct mechanisms for the control of gene expression that operate during the *T. cruzi* life cycle, as the parasite alternates between proliferation and differentiation. The first involves regulation at the level of RNA stability. The second involves an unusual global mechanism of transcriptional control which is associated with structural modifications to the nucleus that affect the replication machinery.

Over the last few years the Frasch group has made a significant contribution to our understanding of the complex family of genes that encode the mucin-like molecules that are the major glycoproteins on the surface of *T. cruzi*. In chapter 3, they summarize the current state of knowledge about the properties and functions of this protective coat and describe in detail some of the elegant experiments that they have carried out to address the post-transcriptional mechanisms by which the parasite regulates expression of this large repertoire of surface antigens. At the metabolic level trypanosomes also differ from their hosts in a number of ways that have importance in terms of pathogenesis. In chapter 4, Wilkinson and Kelly highlight the parasite-specific properties of thiol biochemistry and oxidative defense and discuss the possibility that these distinct enzymes/pathways could provide an opportunity for chemotherapeutic intervention. *T. cruzi* has generally been considered to be deficient in oxidative defense, particularly peroxide metabolism. As these authors point out, this is far from the case and the mechanisms and pathways utilized by the parasite are both sophisticated and flexible.

One characteristic feature of *T. cruzi* is its ability to invade a wide range of mammalian cells. This has major consequences in terms of disease pathogenesis. There is now a large body of evidence that implicates changes to the intracellular concentration of calcium ions (Ca^{2+}) in both host and parasite cells as an important signalling pathway for invasion. In their chapter, Moreno and Docampo discuss some of the complex mechanisms involved in invasion and describe how *T. cruzi* is able to manipulate these signalling systems to induce its own internalization by mammalian cells. However the diverse nature of the *T. cruzi* species and the extensive range of host cells that can be parasitized has made it difficult to identify a single predominant invasion mechanism. To add further to this complicated process, the multigenic and stage-specific nature of many of the *T. cruzi* factors that have been implicated in the invasion process results in yet more complexity. As discussed by Julio Scharfstein for example, parasite proteases that have been shown to generate activation signals for a broad range of host cells are themselves a diverse group of molecules encoded by a large family of developmentally regulated genes. Cruzipain, the major cysteine protease of *T. cruzi* has variously been implicated in immunopathology, host cell invasion, parasite differentiation, and as demonstrated from the authors own work, in the activation of host cell "kinins," small peptides that have diverse roles as mediators of inflammation and circulatory homeostasis. Further studies on the role of these activated kinins in inflammatory responses, along the lines described in chapter 8 may provide new insights into the immunopathology associated with Chagas disease.

Once parasites have been internalized by host macrophages, the major trypanocidal killing mechanism involves the production of nitric oxide (NO) by an inducible NO synthase. However, as outlined by Silva and colleagues, the roles of NO during *T. cruzi* infection are complex and finely balanced. In addition to parasite-killing activity, NO can be involved in suppression of the immune response and there is also considerable evidence for a role in disease pathology. A major feature of Chagas disease is the predominant nonspecific nature of the induced immune response, a phenomenon that has been associated with immunosuppression. This results from the massive polyclonal activation of both B- and T-lymphocytes that occurs early in infection, the vast majority of which do not recognize parasite determinants. Paola Minoprio has played a central role in identifying parasite mitogens involved in this process. This work has provided insights into mechanisms that could mediate polyclonal activation and has focused attention on the potential consequences of this type of response in Chagas disease and other microbial infections. These questions are fully explored in chapter 7, particularly in terms of disease progression and the likely impact on vaccine design strategies.

Gao and Pereira, in the final chapter, introduce the novel concept of "Parasitokines," parasite factors that can mimic the effect of host cytokines, thereby disrupting the immune response and enhancing virulence. In the context of *T. cruzi* infection this can be illustrated by the effects of trans-sialidase (TS), an enzyme that is highly expressed on the surface of the invasive trypomastigote form of the parasite. TS has been shown to have multiple roles during the infection process including its action as a stimulator of T- and B-cell proliferation and immunoglobulin secretion. This polyclonal lymphocyte activation has several downstream consequences that, in combination act to blunt the specificity of the immune response, enhance parasite virulence and drive disease pathology.

In conclusion, rapid technical progress is now providing new approaches to the study of Chagas disease pathogenesis. As should be apparent from the chapters in this book, these advances are occurring at several levels. They are providing novel insights into the genetic and biochemical nature of the parasite, the molecular basis of host:parasite interactions and the varied mechanisms used to avoid immune destruction. This is leading to a greater understanding of disease pathogenesesis and will undoubtedly contribute to the development of new stratagies to alleviate the enormous public health burden that results from Chagas disease.

John M. Kelly, Editor

CHAPTER 1

Genetic Diversity of *Trypanosoma cruzi* and the Epidemiology of Chagas Disease

Michael A. Miles, Matthew Yeo and Michael Gaunt

Abstract

The complex epidemiology of Chagas disease is not fully understood. It has been suggested that distinct genotypes of *Trypanosoma cruzi* may cause the severe (megasyndromes) and benign forms of chronic Chagas disease, which appear to differ in geographical distribution. Multi-locus enzyme electrophoresis (MLEE) and analyses of DNA polymorphisms with several targets have demonstrated that *T. cruzi* has a remarkable degree of genetic diversity. Both isoenzyme and DNA analyses define two major subdivisions within the species, *T. cruzi* I and *T. cruzi* II, with marked heterogeneity and five subdivisions within *T. cruzi* II (II a-e). Population genetic analyses have indicated that *T. cruzi* is predominantly clonal, although the isolates studied have mainly been sporadically collected over vast geographical distances. However, *T. cruzi* IId and IIe display putatively hybrid phenotypes in the form of multiple heterozygous isoenzyme phenotypes. Phylogenetics analyses have confirmed the hybrid nature of *T. cruzi* IId and IIe and indicate that genetic exchange has contributed to the evolution of genetic diversity in *T. cruzi.* The *T. cruzi* strain selected for the genome sequencing project is a hybrid (IIe) strain. We have proved experimentally that *T. cruzi* I has an active capacity for genetic exchange using parental isolates taken from a single locality where parents and hybrids were sympatric. Experimentally derived hybrid clones displayed a combination of parental phenotypes and genotypes, indicating that *T. cruzi* may evolve via hybridization, aneuploidy and genome erosion. We propose tentative associations for *T. cruzi* I, with the maruspial *Didelphis* (common opossum), the triatomine genus *Rhodnius* and the palm tree ecotope, and for *T. cruzi* II with edentates (armadillos), rodents, the triatomine genus *Triatoma* and the terrestrial ecotope. We conclude that there must be a link between *T. cruzi* genotype and outcome of infection but the nature of the link in terms of disease pathogenesis remains to be defined.

Introduction

The protozoan parasite *Trypanosoma cruzi,* the agent of Chagas disease, is considered to be the most important parasite in Latin America, causing morbidity exceeding that due to malaria. Serological surveys suggest that up to 20 million people carry *T. cruzi* infection. Many millions are still exposed to infection, despite the success of recent control programmes. Chagas disease is a complex zoonosis with an epidemiology that is not fully understood.[1]

Transmission of *T. cruzi* to the mammal host is usually by contamination with infected faeces of the insect vector, the triatomine bug (Hemiptera: Reduviidae). Triatomine bugs acquire their infection through feeding on an infected mammal. The infective forms (metacyclic

Molecular Mechanisms in the Pathogenesis of Chagas Disease, edited by John M. Kelly.

trypomastigotes) in bug faeces can penetrate mucous membranes or abraded skin. Transmission may also be by blood transfusion, by organ transplantation, congenital, or by eating infected raw meat or food contaminated with infected bug faeces.

In the mammal host *T. cruzi* is primarily intracellular. Trypomastigotes do not multiply in the blood: they invade non-phagocytic or phagocytic cells, and divide by binary fission as small ovoid amastigotes (without a flagellum) to produce a pseudocyst. Pseudocysts may form in many tissue types but are particularly common in muscle, especially heart and smooth muscle. The pseudocyst ruptures after 5 days or more releasing new trypomastigotes, some to re-invade cells and others to circulate in the blood.

T. cruzi is confined to the Americas. More than 150 species of mammal, of 24 families, have been reported as infected and all mammals are thought to be susceptible to infection. Birds and reptiles do not carry *T. cruzi*. Approximately 133 species of triatomine bug have been described. The vast majority of triatomine species occur only in the New World. Only 13 species occur in the Old World, 8 of which are related to *Triatoma rubrofasciata*, which has spread to ports around the world with its vertebrate host, the rat (*Rattus rattus*). The remaining 5 Old World species of triatomine belong to the unusual Indian genus *Linshcosteus*. The natural habitats of triatomine bugs are silvatic, for example in palm trees, burrows, hollow trees or among rocks. Endemic Chagas disease is transmitted by the few triatomine bugs species that have adapted to colonise and thrive in human dwellings.[2] The most notorious domiciliated triatomine bug species (Fig. 1) are: *Triatoma infestans*, which is the main vector in southern South America; *Rhodnius prolixus*, in northern South America and Central America; *Panstrongylus megistus*, in eastern and central Brazil; *Triatoma brasiliensis*, in north eastern Brazil, and *Triatoma dimidiata*, also in northern South America and Central America. Interestingly, the type of household infestation may reflect the natural habitat of the bug species concerned. Thus *T. infestans*, found naturally among rocks, can be found in quite good quality housing, and even in tile roofs. *R. prolixus*, a palm-dwelling species, is abundant in palm roofs. *P. megistus*, which inhabits tree-holes and burrows, is abundant in timber-frame and mud-walled houses. Despite these subtleties, bug infestation is essentially associated with poor quality housing, and Chagas disease is fundamentally a disease of poverty. The risk of acquiring Chagas disease outside Latin America is low. The *T. cruzi* zoonosis extends well into North America, bugs occasionally invade the area around the house when there are peridomestic dogs or opossums, yet vector-borne human Chagas disease is very rare in the USA. Transmission of *T. cruzi* by blood transfusion or by organ donation outside the usual endemic areas is known and is a continuing risk with increasing migration of human populations from Latin America. Ideally all migrants who were resident in rural regions infested by domiciliated triatomines should be screened serologically for antibodies to *T. cruzi* and if seropositive they should be excluded from acting as donors.

Organisms closely related to *T. cruzi*, and of the same subgenus (*Schizotrypanum*) are cosmopolitan in bats. The relationship between *T. cruzi* and bat schizotrypanosomes is not clear. *T. cruzi* may be more ancient than bat schizotrypanosomes, or may be derived from them. The other South American human trypanosomiasis, due to *Trypanosoma rangeli*, is transmitted by triatomine bugs of the genus *Rhodnius*, in Central America and rarely in South America. Although morphologically distinct, *T. rangeli* may be closely related to *T. cruzi*. *T. rangeli* is considered to be non-pathogenic to humans. However, it can be pathogenic to triatomine bugs, in which it migrates to the salivary glands to be transmitted by the bite and not by contamination with faeces. A third trypanosome species, *Trypanosoma conorhini*, is transmitted by *Triatoma rubrofasciata* to rats, by the contaminative route. The other kinetoplastid parasites responsible for human diseases, namely the African trypanosomes (*Trypanosoma brucei*) and *Leishmania*, are thought to have diverged from *T. cruzi* long ago; *T. cruzi* is considered to be more closely related to *Leishmania* than *T. brucei*.

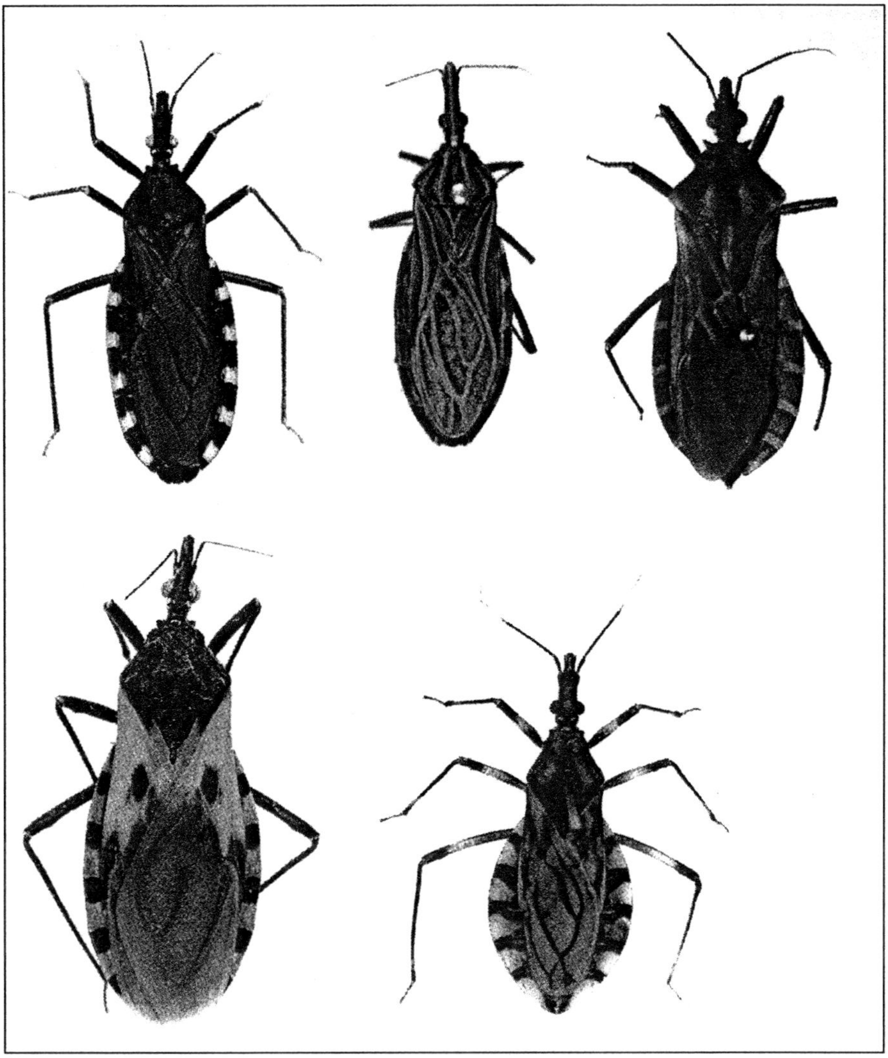

Figure 1. The principal domiciliated triatomine vectors of *Trypanosoma cruzi*. From left to right; (above) *Triatoma infestans; Rhodnius prolixus; Panstrongylus megistus;* (below) *Triatoma dimidiata; Triatoma brasiliensis.* See text for geographical distributions.

Initial, acute phase human infection with *T. cruzi* may be fatal in approximately 10% of cases, particularly in children. Once acquired, infection is usually carried for life, unless parasites are eliminated by drug treatment. Chagas disease may have a devastating effect on the human host. The chronic disease is primarily a heart disease, causing ECG abnormalities and cardiomyopathy in up to 30% of those infected (Fig. 2). A smaller proportion of infected individuals develop enlargement and dysfunction of the oesophagus or colon (megaoesophagus; megacolon). These signs of chronic Chagas disease usually arise years after the initial infection. Meningoencephalitis may also occur, particularly in congenital cases and in AIDS-associated reactivation of chronic infection. The pathogenesis of Chagas disease is only partially understood.

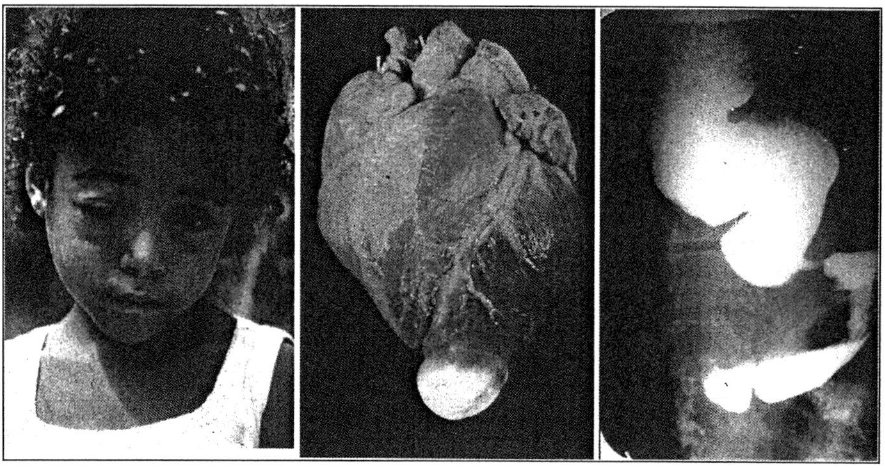

Figure 2. From left to right: Romana's sign (unilateral conjunctivitis and oedema) in acute Chagas disease; apical aneurysm seen post mortem in chronic Chagas disease (courtesy of Joao Oliveira); severe megaoesophagus seen on X-ray in chronic Chagas disease (courtesy of Joao Oliveira).

The roles of autoimmunity and of persistent low level infection are controversial.[1] It is clear that neurone loss, particularly from the parasympathetic system, may occur during the acute phase of infection. The loss is exacerbated by age, leading to sympathetic dominance and heart disease, or failure of peristalsis in the alimentary tract. In some patients chronic inflammation and progressive heart disease might be triggered by autoimmunity. Treatment, with benznidazole (Roche) is recommended for all acute cases (including congenital cases and AIDS related reactivation of acute infection) and for chronic cases in children. Long-term therapy is required, side-effects may be severe and the parasite is not always eliminated.[1]

The geographical distribution of chronic Chagas disease is enigmatic. There appears to be marked regional differences in the prevalence of severe chronic symptoms. In particular, megaoesophagus and megacolon are well known in central and eastern Brazil but virtually unknown in northern South America and Central America. There also appears to be marked regional variation in response to chemotherapy. One proposed explanation of these differences is that *T. cruzi* is not a single entity but a heterogeneous complex with diverse biological characteristics. More explicitly, it is suggested that there are at least two distinct genotypes, one causing severe chronic Chagas disease and the other responsible for more benign disease.[1]

Heterogeneity of *T. cruzi*

The concept that differences in clinical presentation and success of chemotherapy for Chagas disease might reflect diversity of *T. cruzi* was supported by other early observations. Thus *T. cruzi* strains appear to differ in virulence and histotropism, in infectivity to different triatomine species, and in antigenic composition. There was no reliable intrinsic method for resolving the potential heterogeneity of *T. cruzi* until the 1970s when multi-locus enzyme electrophoresis (MLEE) allowed a systematic comparison of phenotypic and, by interpretation, genotypic differences between *T. cruzi* isolates. A classic field and laboratory investigation published in 1977[3] transformed our understanding of *T. cruzi*. Field isolates from houses infested by *P. megistus*, in Sao Felipe, Bahia State, Brazil, differed radically from silvatic isolates recovered from the common opossum (*Didelphis albiventris*). This observation led to the finding of a silvatic triatomine bug species (*Triatoma tibiamaculata*) in opossum refuges in bromeliad epiphytes and the description of separate domestic and silvatic transmission cycles of *T. cruzi* in

Sao Felipe.[4] The domestic and silvatic *T. cruzi* strains were so distinct that 11 out of 18 enzymes separated them, a level of distinction much greater than that separating biologically and clinically defined species of *Leishmania*. The distinct *T. cruzi* strain groups were named zymodeme 1 (Z1) and zymodeme 2 (Z2). The initial description of Z1 as silvatic and Z2 as domestic in Sao Felipe did not imply a general attribution of Z1 to silvatic cycles and Z2 to domestic cycles, although this misconception sometimes appears in the scientific literature. In fact as early as 1978[5] Z1 was isolated from an acute case of Chagas disease in Amazonian Brazil. In 1981,[6] a comparative study of 316 Venezuelan and Brazilian *T. cruzi* isolates demonstrated that, in contrast to Brazil, Z1 was circulating in domestic transmission cycles in Venezuela. This led to the suggestion that Z2 might be responsible for the more severe chronic Chagas disease in Brazil, and Z1 for the more benign disease in Venezuela.

A third uncommon group of *T. cruzi*, with uncertain affinities and named Z3,[5] was initially isolated from a human acute case in Amazonian Brazil but also turned up sporadically elsewhere in Brazil, Venezuela and other countries[6] (see Transmission cycles, below). A variant of Z3, also isolated from Amazonian Brazil was designated Z3/Z1ASAT[7] (ASAT referring to the enzyme aspartate aminotransferase).

Subsequently MLEE has been applied widely but sporadically to the analysis of *T. cruzi* isolates from several other countries in Latin America. Isoenzyme analysis of *T. cruzi* isolates from Bolivia and Chile produced a strange finding in that isolates with prominent multiple isoenzyme patterns were very abundant.[8,9] These patterns were typical for those expected of a diploid organism with heterozygous alleles at a single locus. Heterozygous patterns for enzymes predicted to be dimeric, such as glucose phosphate isomerase (GPI), were triple-banded, with the three bands equidistantly separated and the central band most intense. Heterozygous patterns for enzymes predicted to be monomeric, such as phosphoglucomutase (PGM) were double-banded. Multiple patterns were retained in biological clones of *T. cruzi,* so they could not be due to mixed infections. Estimates of the sub-unit numbers for several *T. cruzi* enzymes were obtained by biochemical determination of the enzyme molecular sizes. The sub-unit numbers obtained were as suspected from interpretation of the isoenzyme profiles.[10] Because some of the multiple isoenzyme bands had similar mobilities to those of Z2 this new group was named Bolivian Z2. Heterozygous profiles were also common in *T. cruzi* isolates from Paraguay (Paraguayan Z2) but they were subtly different from Bolivian Z2.[11] The multiple heterozygous loci in such isolates suggested that they might be derived from a hybridization event between two different *T. cruzi* strains. Examples of heterozygous or "hybrid" isoenzyme profiles are shown in Figure 3.

Thus isoenzyme analysis yielded two major groups, Z1 and Z2, and at least four others Z3; Z1/Z3ASAT; Bolivian Z2 and Paraguayan Z2. The more extensive study of Tibayrenc and his collaborators, using 15 enzymes, defined 43 different zymodemes. Nevertheless phylogenetic analyses still consolidated these zymodemes into two major groups, one group (zymodemes 1-25) corresponding with Z1, and a second (zymodemes 26-43) incorporating Z2 and the four lesser groups.[12]

In parallel with MLEE, restriction fragment length polymorphism (RFLP) analysis of kinetoplast DNA (kDNA, schizodeme analysis) was developed for typing *T. cruzi* isolates.[13] Thereafter several other molecular methods were introduced, including random amplification of polymorphic DNA (RAPD),[14] comparison of ribosomal and mini-exon DNA sequence polymorphisms,[15] and microsatellite analysis.[16] The data from these methods are superficially complex, with some confusion of terminology. In reality, however, the picture that emerges is consistent and coherent.

Schizodeme analysis correlated well with MLEE and readily distinguished Z1 from Z2.[17] RAPD analysis also correlated well with isoenzyme analysis giving similar major divisions and sub-groups. The ribosomal and mini-exon DNA sequence polymorphisms were less sensitive

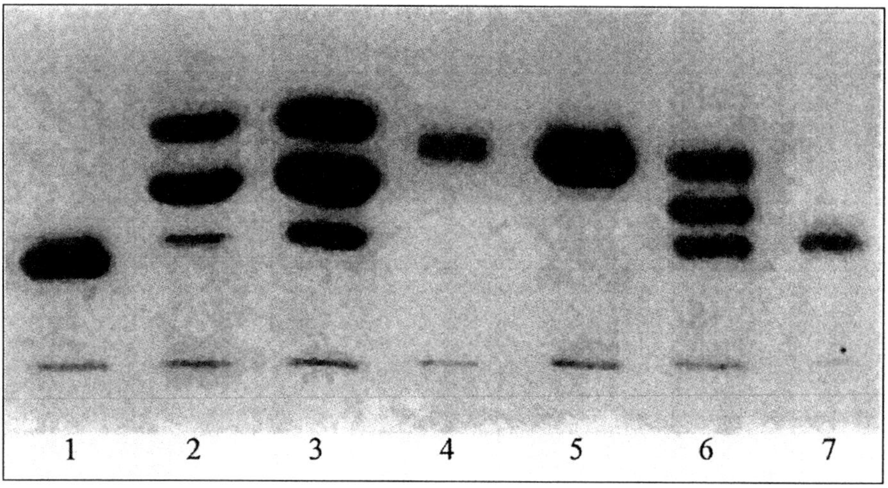

Figure 3. Starch-gel electrophoresis of *T. cruzi* glucosephosphate isomerase, tracks (left to right): 1, *T. cruzi* I (Z1); 2, 3 *T. cruzi* IId (Bolivian Z2); 4, 5 *T. cruzi* IIb (Z2); 6 *T. cruzi* IIe (Paraguayan Z2); 7, *T. cruzi* IIa (Z 3).

than isoenzyme and schizodeme analysis. Nevertheless both methods were informative and split *T. cruzi* into major sub-divisions, supportive of the isoenzyme, schizodeme and RAPD groups.[18] The 100 base pairs (bp) at the 3' end of the 24Sα ribosomal RNA gene proved to be dimorphic. PCR amplification yielded products of either 125bp or 110bp, with a few isolates giving both products. Similarly amplification of hypervariable intergenic regions in the mini-exon gene array initially gave two sub-specific groups of *T. cruzi*. There was a precise correlation between the 24Sα and the mini-exon groups. Furthermore those isolates showing a 24Sα product of 110bp corresponded with Z1, and those isolates giving a 24Sα product of 125 bp or both 110bp and 125bp encompassed Z2. Z3 isolates gave a mini-exon non-transcribed spacer product carrying an unusual insertion.[19] RFLP analysis of two further regions, the internal transcribed spacer between the 18S rRNA and the S3 rRNA (ITS1) genes and between the S3 rRNA and the 24Sα rRNA (ITS2) genes also supported sub-division of *T. cruzi* into two major phylogenetic groups, and further divided Z3 isolates into two sub-groups.[20] A separate study of North American isolates by RFLP analysis of the 18S rRNA gene (riboprinting) defined 3 "ribodemes".[21]

Finally PCR amplification of microsatellite alleles (CA repeats) has been applied to study diversity in *T. cruzi*. Phylogenetic analysis, assuming a step-wise mutation model for the microsatellites, separated two groups, one corresponding with Z1 (24Sα 110bp) and the second encompassing Z2 (24Sα 110bp or 110bp + 125bp) but with marked heterogeneity once again among the group that included Z2.[16]

Thus these isoenzyme and DNA based studies of *T. cruzi* diversity consistently defined two major sub-divisions within the species, albeit with marked heterogeneity and sub-groups within one of them. In an attempt to remove confusions from contradictory zymodeme and lineage terminologies the two major sub-divisions were redefined in 1999 by international consensus as *T. cruzi* I and *T. cruzi* II.[22] *T. cruzi* I corresponds with Z1 and *T. cruzi* II incorporates Z2 and other groups.

A recent detailed comparison of MLEE, RAPD, 24Sα rRNA gene, non-transcribed mini-exon spacer, and RFLP analysis of the 18S rRNA gene has revealed a surprisingly simple and uniform picture. This comparison once again defines two major sub-divisions, *T. cruzi* I and *T.*

Table 1.* T. cruzi *subspecific groups

Subspecific Designation	Zymodeme (Miles)	Zymodemes (Tibayrenc)	Example of Reference Strain
T. cruzi I	Z1	1-25	X10 clone 1: *Tibayrenc* and Miles, 1983[8] WA250: Miles et al, 1977[3]
T. cruzi IIa	Z3	26-29	Can III clone 1: Miles et al, 1978[5]
T. cruzi IIb	Z2	30-34	Esmeraldo clone 3: Miles et al, 1977[3]
T. cruzi IIc	Z3/Z1 ASAT	35-37	M5631: Miles et al, 1981[6] X9/3; X109/2 : Chapman et al, 1984[10]
T. cruzi IId	Bolivian Z2 "heterozygous"	38-39	SC43; 92:80: Tibayrenc and Miles, 1983[8]
T. cruzi IIe	Paraguayan Z2 "heterozygous"	40-43	X57/3; P69/8: Chapman et al, 1984[10] (and CL Brener—the genome project strain)

cruzi II. *T. cruzi* II is divided into 5 sub-groups IIa-IIe.[12] Remarkably all of the five sub-groups correspond with groups previously designated by early work using isoenzyme electrophoresis (Table 1). This picture provides a working framework for future studies of *T. cruzi* diversity. Nevertheless it is certain to be modified as more DNA sequence and phylogenetic analyses emerge. Note that this summary places Z3 and Z3/Z1 ASAT within *T. cruzi* II as groups *T. cruzi* IIa and IIc respectively. Other authors have considered these two groups to have closer affinities with *T. cruzi* I (Z1). Concisely stated this large body of data leads to the conclusion that *T. cruzi* I corresponds with Z1, IIa with Z3, IIc with Z3/Z1 ASAT, IIb with Z2, IId with Bolivian Z2, and IIe with Paraguayan Z2/CL Brener (Table 1).

A significant weakness of all these studies of *T. cruzi* heterogeneity is that the isolates involved are sporadically collected across vast geographical distances. There have been few intensive studies of single localities and single transmission cycles. The lack of samples has to some extent restricted population genetic analyses, although the latter have been informative.

Population Genetics

Population genetics measures gene flow between or within populations. The Hardy-Weinberg equilibrium test looks at the proportions of the different possible allele types in a population to assess whether there is random re-assortment (segregation) of different alleles at a given locus, as a measure of random mating (panmixia) or as a measure of restriction of random mating. Ideally multiple loci should be examined, and with many samples from the population being studied. Ploidy must be greater than one and should be known. *T. cruzi* has generally been presumed to be diploid in population genetic analyses. Karyotype analysis by pulsed field gradient gel electrophoresis has indicated that *T. cruzi* is at least diploid.[23]

Alternative multi-locus population genetic analyses use linkage disequilibrium tests. They do not demand that ploidy be known. Such tests look for departure (linkage disequilibrium) from random reassortment (panmixia) among independent loci. As pointed out by Gibson and Stevens, in their excellent review,[23] these tests are fundamentally dependent upon the extent and richness of the available data. If small samples sizes and limited data are available, such tests are vulnerable and likely to lead to biased and distorted outputs.

A series of population genetic analyses of *T. cruzi* isolates in the 1980s and 1990s consistently indicated that *T. cruzi* was sub-structured into asexual clonal populations. It was initially concluded, prematurely, that genetic exchange was absent from *T. cruzi*.[24] The same results were obtained irrespective of whether they were based on isoenzymes, RAPD analysis or RFLP analysis. A similar conclusion was reached using microsatellites.[16] Although these studies strongly indicate the predominant clonality of *T. cruzi* they are vulnerable to the lack of data sets from sympatric isolates and single transmission cycles of *T. cruzi*. Importantly, the perception of genetic structure in *T. cruzi* has shifted within the last decade from the overwhelming emphasis on clonality and absence of genetic exchange to acceptance that genetic exchange may have made a contribution to the evolution of *T. cruzi*. It is interesting that this shift in perception parallels a similar realization that genetic recombination makes a significant contribution to the diversity of bacterial populations.[25] As we will demonstrate here, in addition to circumstantial population genetics evidence and new phylogenetic evidence, there is now experimental proof that *T. cruzi* has an extent capacity for genetic exchange. The implications of a capacity for genetic exchange may be profound. *T. cruzi* with new characteristics may arise from genetic exchange events, and the new phenotypes may be highly competitive in existing or new ecological niches.

Genetic Exchange in *T. cruzi*

The discovery of the putatively hybrid phenotypes of Bolivian Z2[8,9] and Paraguayan Z2[11] suggested genetic exchange had at least contributed in the past to the genetic diversity of *T. cruzi*, even though genetic exchange might not be occurring in present populations of *T. cruzi*. The hybrid strains are mainly found in southern latitudes or mountainous regions, in the southern cone countries (see below) of South America. This led to the suggestion that the multiple isoenzyme bands might confer selective advantage in these geographical regions. Perhaps the distinct isoenzymes allowed versatile enzyme activity over a wide range of temperatures, to which triatomine bugs and *T. cruzi* might be exposed during different seasons. The GPI isoenzymes have different temperature stabilities. However, GPI kinetic parameters did not differ between isolates with single and triple banded GPI phenotypes.[26] Thus, the hypothesis that a multiple heterozygous phenotype confers a selective advantage could not be confirmed, although it has yet to be fully explored.

In an important recent development Machado and Ayala[27] have used phylogenetic tests to re-examine relationships between *T. cruzi* I and *T. cruzi* II, including sub-groups IId and IIe that have hybrid phenotypes. DNA sequence was determined for segments of three independent loci, namely the enzymes trypanothione reductase (TR), dihydrofolate reductase-thymidylate synthase (DHFR-TS) and a region spanning cytochrome oxidase subunit II (COII)/NADH dehydrogenase subunit 1 (ND1) of the mitochondrial maxicircle. If the populations are asexual and clonal the genes should share their history of divergence, whereas if genes have different histories of divergence (incongruent gene genealogies), the implication is that genetic exchange has occurred. Intragenic recombination was not detected by a test to detect decay of linkage disequilibrium with distance. However, maximum likelihood phylogenetic analysis showed three gene geneaologies that were incongruent, contrary to the presumption of clonality. Strikingly, the nuclear genomes of *T. cruzi* isolates representative of groups IId and IIe were confirmed as hybrid. Putative parental haplotypes were similar or identical to those of groups *T. cruzi* IIb and IIc. Note that the simple isoenzyme phenotype published almost two decades ago (Fig. 3) shows that the GPI bands for IIb and IIa correspond with the two outer bands present in the hybrid strain IIe.[28] Note also the differences between the multiple bands of IId and IIe. Overall the implications from this landmark study are that multiple genetic exchange events have contributed to the genetic diversity and evolution of *T. cruzi*, both across subgroups of *T. cruzi* II, and within them.[27] The CL Brener clone of *T. cruzi* that is the focus of the genome sequencing project belongs to the hybrid group IIe. This has significant implications for the workload of that project and for the interpretation of its output. If CL Brener has a high DNA content and in part has triple ploidy or greater, more sequences must be determined, and they must be assembled with caution. The outcome may be partial insight into the sequence of both progenitor strains, which might better have been achieved by sequencing one, less complex example of *T. cruzi* II and one example of *T. cruzi* I.

Proof of an extant capacity for genetic exchange in *T. cruzi* I has come from our own experimental work on *T. cruzi* hybrids. In 1996 we published the discovery of putative parental and hybrid phosphoglucomutase (PGM) phenotypes circulating sympatrically in a transmission cycle of *T. cruzi* I in the Amazon forest at the locality of Serra das Carajas, Brazil.[29] (We had previously reported heterozygous phenotypes and at least one of the putative corresponding homozygous phenotypes circulating sympatrically in Paraguay[11]).

The frequencies of the PGM alleles in Serra das Carajas appeared to be in Hardy Weinberg equilibrium, although this data set was too small to be statistically significant. By analogy the first evidence for genetic exchange in *Trypanosoma brucei* was initially based on patterns and frequencies of isoenzyme phenotypes among a small number of field isolates.[30] Genetic exchange in *T. brucei* was subsequently proved by laboratory crosses.[23] Only recently has a larger series of field isolates been analysed by minisatellites, confirming the relevance of recombination to the epidemiology of African trypanosomiasis.[31]

As summarized in Stothard et al,[32] we reported the experimental recovery of double drug resistant hybrids from copassage of genetically transformed parental pairs carrying single drug resistant markers. Biological clones of the putative parents were transformed to be resistant to either hygromycin (150mg/ml) or G418 (120mg/ml). These resulting clones were passaged singly or together, axenically, through mammalian cells in vitro, and through triatomine bugs and mice in vivo. Resultant populations were selected for growth in the presence of both drugs.

Data from these experiments are still being collected and in due course the details will be published. Nevertheless we affirm here that biological clones derived from the double drug-resistant population of *T. cruzi* I have the following characteristics:

- The hybrid clones carry episomal constructs derived from both parents, as demonstrated by PCR analysis;
- The hybrid clones have a combination of the parental PGM phenotypes;
- By karyotype analysis the hybrid clones have both parental cysteine protease (CP) genotypes;
- RAPD analysis demonstrates band sharing between the hybrid clones and each parent.
- DNA sequencing indicates uniparental inheritance of maxicircle kinetoplast DNA.

Moreover, there are strong genetic parallels between the outcome of this breeding experiment and characteristics of isolates of *T. cruzi* originating from the field. Further molecular analyses suggest aneuploidy and genome erosion in the progeny of this experiment. We have thus demonstrated unequivocally genetic hybridization in *T. cruzi* I. As far as we are aware this is the first and only experimental production of *T. cruzi* hybrids. It is interesting to recall, however, that the naturally occurring, multiple heterozygous ("hybrid") phenotypes feature in *T. cruzi* II (IId and IIe).

We conclude that genetic hybridization in *T. cruzi* is not simply historical, but is a significant and active mechanism generating genetic diversity. This may play a major role in the generation of recombinant genotypes with enhanced vigour and in the spread of virulence characteristics or drug resistance.

Several years ago Dvorak and his collaborators put forward the suggestion that *T. cruzi* genetic diversity is generated by hybridization events.[33] In meticulous work, they quantitated the DNA in *T. cruzi* I (Z1) and *T. cruzi* II (Z2 and Z3) by mithramycin staining and flow cytometry (FACS).[34] The wide range in DNA content indicated vast genetic diversity in *T. cruzi*. The knowledge that there were hybrid phenotypes led to the suggestion that *T. cruzi* evolved by hybridization, aneuploidy, and genome erosion, analogous to the way in which some plants generate diversity.[33] Unpublished observations (S. Obado and J. Kelly, personal communication) strongly support both aneuploidy and gene erosion in the CL Brener strain of *T. cruzi*.

Multiclonality

As a brief aside we draw attention to the potential of plating of *T. cruzi* primary isolates to obtain biological clones and to assess diversity in naturally infected triatomine bugs and mammals. We have modified the plating method of Mondragon et al,[35] and shown that it can be applied directly to infected triatomine bug faeces. We have applied this procedure to triatomine bugs from geographically distant locations. Different bugs from a single locality may carry genetically distinct *T. cruzi* clones. Mixed genotypes may also be isolated from single triatomine bugs (Fig. 4). Further details of this method and its applications will be published elsewhere (Yeo et. al., in preparation).

Transmission Cycles and Host Associations

T. cruzi I predominates in enzootic cycles in the vast Amazon basin and in domestic transmission cycles in all the endemic countries that lie to the north of the Amazon basin. *T. cruzi* I is also found in some silvatic transmission cycles south of the Amazon basin.[2,3]

Enzootic *T. cruzi* in the Amazon basin is notoriously complex, with at least 22 mammal species, excluding bats, recorded as infected, although in some of these mammals prevalence may be very low. Ten Amazonian triatomine species have been reported as infected with *T. cruzi*. The most common naturally infected host seems to be the common opossum, *Didelphis*. In fact, *T. cruzi* I has been associated with *Didelphis* repeatedly over a vast geographical range. We have therefore proposed that *T. cruzi* I has an ancient evolutionary history in association with *Didelphis*-like marsupials. Interestingly, anal gland infections of *T. cruzi* I are known to occur in *Didelphis*, with morphological forms similar to those seen in the insect vector. It is not clear whether this is a primitive life cycle, which does not require vector involvement, or an aberrant occurrence. It is generally thought that mammalian trypanosomes have evolved from monoxenous kinetoplastids of insects.

T. cruzi II is the predominant cause of Chagas disease throughout the southern cone countries of South America (Argentina, Bolivia, Brazil, Chile, Paraguay, and (presumably) Uruguay). It is throughout this region that clinical Chagas disease appears to be more severe and megasyndromes most common. The vector in this region is *T. infestans*, which is thought to have spread from a silvatic habitat in Bolivia, associated with guinea pigs. We have already noted (above) the strange geographical distribution of *T. cruzi* IId and IIe in the high altitude western flank of South America and the southern regions of the continent. The natural reservoir hosts of *T. cruzi* II are incompletely known, although it has been reported from marmosets (*Leontopithecus rosalia*) and an opossum species (*Philander opossum*) in the vicinity of Rio de Janeiro.[36] Briones et al[37] have proposed that *T. cruzi* II has an origin in a North American placental mammal host. We think that this is unlikely. Rodents and primates are believed to have arrived in South America 40 million years ago (mya), which is 25 million years later than marsupials (65 mya).[2] We have proposed that *T. cruzi* II has an association with armadillos (edentates), which were present in South America as early as marsupials.[2] We have further proposed that transfer of *T. cruzi* II between armadillos and rodents may have occurred secondarily

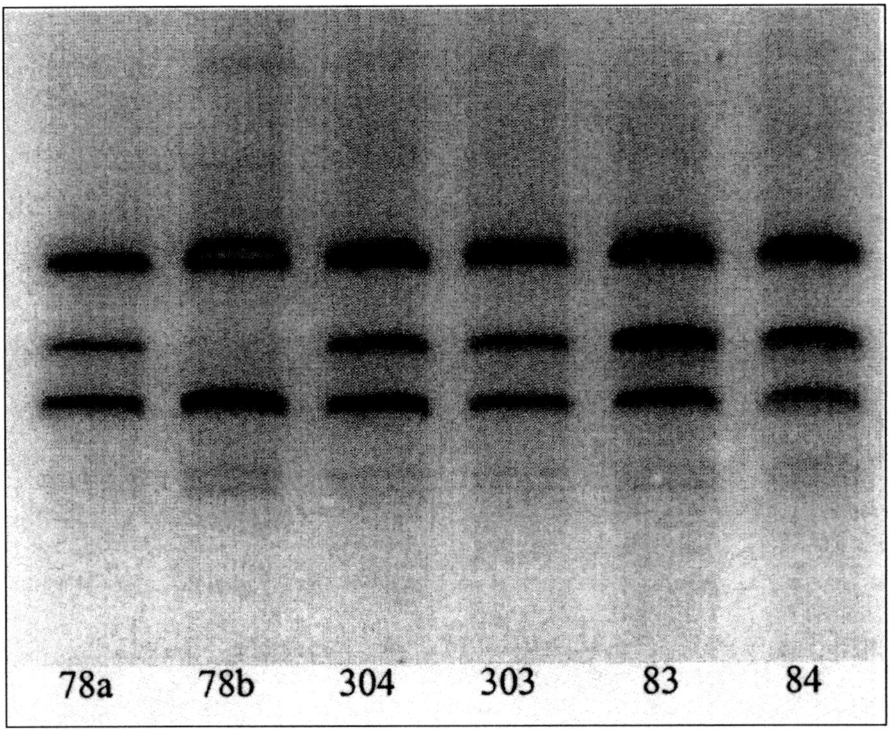

Figure 4. RAPD profiles (generated with Stratagene 2000 *Taq* polymerase) of individual biological clones of *T. cruzi* isolated and cloned directly from naturally infected triatomine bugs by plating on agar (Yeo and Miles, unpublished). This Figure illustrates our observation that distinct *T. cruzi* genotypes may be found in different bugs from the same locality and distinct genotypes may also be found in a single bug (78). Multiple clones of only one genotype were found in bugs 304, 303, 83 and 84; multiple clones of each of the two genotypes were found in bug 78. Results were highly reproducible on repeat examination.

through the sharing of terrestrial habitats, that is burrows or rock piles.[2] The suggestion that *T. cruzi* II has an ancient association with edentates accords with older estimates for the age of divergence between *T. cruzi* I and *T. cruzi* II but not with estimates that give that event a much more recent date.[37,27]

We have tentatively expanded the hypothesis that *T. cruzi* has an ancient association with *Didelphis* by proposing more specifically that it evolved in the palm tree habitat with the triatomine genus *Rhodnius*, virtually all species of which are found in palm trees. Exceptions are *Rhodnius domesticus*, found in bromeliads and hollow trees, and *Rhodnius paraensis*, an entirely valid distinct species, which was discovered in a tree hole refuge of the arboreal spiny rat *Echimys chrysurus* but has never been found since. By analogy, we have suggested that *T. cruzi* II has evolved in a terrestrial habitat, particularly with the triatomine genus *Triatoma*. Note that at least 20 *Triatoma* species are associated with terrestrial rocky habitats or burrows. The age of triatomine bugs is a contentious issue. We date *Rhodnius* back to the emergence of palms, other authors consider triatomines to be much more recent.

T. rangeli is also strongly associated with *Didelphis* and may have evolved similarly in the palm tree habitat. There is heterogeneity within the species *T. rangeli*. The precise relationship between *T. cruzi* and *T. rangeli* is not clear: they appear to be closely related on molecular evidence, even though they are within two distinct sub-genera on morphological grounds. We have proposed that bat trypanosomes are derived from *T. cruzi* by habitat sharing and have been spread around the globe with the dispersion of bat species by flight. This is by no means certain. See Gaunt and Miles[2] for more details of these arguments.

T. cruzi Genotypes and Clinical Prognosis

The possible link between the *T. cruzi* genotypes carried by a patient and the clinical outcome of infection is a topic of great interest. A link between infecting genotype and clinical prognosis lacks formal proof. The idea that *T. cruzi* I is more benign and *T. cruzi* II more severe is appealing based on the distribution of *T. cruzi* II in southern endemic areas of Chagas disease.[6] Nevertheless, given the diversity of *T. cruzi* II this is a gross simplification. What, for example, is the difference between the virulence and pathogenicity of *T. cruzi* IIb (Z2 as originally described) and the hybrid strains *T. cruzi* IId and IIe? Nevertheless it is difficult to deny a link between *T. cruzi* genotype and clinical outcome and there is experimental evidence to support such a relationship.

Dvorak, in careful experimental studies of *T. cruzi* infection in vitro and in mice demonstrated an association between genotype and growth rate and genotype and course of infection in the vertebrate host, and a correlation with drug sensitivity.[38] Tibayrenc and his collaborators have reported similar results.[12] It is difficult to extrapolate the conclusions of these studies directly to human Chagas disease. The obvious way of addressing this question is to genotype *T. cruzi* isolates from a large number of patients with a diverse range of defined clinical pictures. This is not a straightforward investigation. It is known that some patients carry multiple clones and not all clones may be detected. Furthermore, the picture of infection that emerges at the time of isolation may be very different from that at the time of the acute phase of infection, when most of the histological damage is thought to occur. Although Luquetti et al,[39] found that *T. cruzi* II was consistently isolated from chronic Chagas disease in Goias state, Brazil, both *T. cruzi* I (Z1) and *T. cruzi* II (Z2) were associated with similar clinical pictures in the acute phase of infection. Mixed acute phase infections could not be excluded, with early elimination of *T. cruzi* I (Z1) and retention of *T. cruzi* II (Z2). Whilst presence and absence of obvious signs of chronic Chagas disease, such as apical aneurysm of the left ventricle or megaesophagus, are easy to score, it is notoriously difficult to define clearly the full clinical picture in chronic Chagas disease and to classify cases as truly asymptomatic. Finally there are multiple confounders to be excluded during such an investigation, including co-infection, autoimmunity precipitated by other causes, genetic predisposition of some human genotypes, nutritional, behavioral and other environmental factors. Nevertheless reassessment of the stated low prevalence prevalence of megasyndromes in central and northern South America as compared to the southern cone would be a valuable contribution.

A fascinating clonal-histotropic model of differential pathogenicity has been put forward.[40] This suggests, as indicated by previous studies in vitro, that *T. cruzi* clones have differing predilection for distinct cell types. In a multi-clonal infection it is suggested that some clones are more abundant in certain organs. Although data are limited, this principle has been confirmed by kDNA signature typing the *T. cruzi* isolates in organ biopsies from chronic cases of Chagas disease. The variable kDNA regions are amplified by PCR, purified and then subjected to a second round PCR, with a low stringency, single specific primer (LSSP-PCR). Results indicated a differential distribution of kDNA signatures between heart and oesophagus.[40]

One theoretical driving force for research on this topic is the speculation that individuals carrying particularly pathogenic *T. cruzi* genotypes could be identified reliably and cost effectively and given more intense clinical follow-up and preferential access to clinical management. The 20 million or so individuals carrying *T. cruzi* might be classified by serological tests into high risk or low risk of megasyndromes or chagasic heart disease.

On the basis of reviewing the evidence we must conclude that there is indeed a link between *T. cruzi* genotype and clinical outcome of infection although the nature of the link and the mechanisms of pathogenesis have not been defined.

Future Work

To conclude we will make a few, rather obvious comments on areas of future research. Firstly it is essential to perform more thorough studies of the genetic diversity of *T. cruzi* and representative transmission cycles at single localities and with much larger data sets. There has been a tendency, albeit fruitful, to re-examine available *T. cruzi* isolates, to some extent producing a reiteration of former observations, rather than to obtain adequate new sets of field isolates. Secondly, further DNA sequence based analyses should be applied to such samples, alongside reference strains.[27,41] Thirdly, the way forward is now open to explore the mechanisms of genetic hybridization and recombination in *T. cruzi* experimentally. Fourthly, the genome sequencing project should be re-orientated to provide data that can underpin the epidemiological studies; ideally the genome sequence should be determined for at least the *T. cruzi* I and *T. cruzi* II genomes.[42] Availability of these genome sequences will inevitably lead to comparative functional analysis and insight into histotropisms and mechanisms of pathogenesis. Finally, advances in phylogenetics and molecular dating, together with better understanding of parasite-host and parasite-vector relationships is likely to yield intriguing insight into the evolutionary history of these organisms. An unequivocal comparison of parasite genotypes, human genotypes and clinical outcomes seems a distant aim, requiring further technical advances and huge resources.

Acknowledgements

We thank the Wellcome Trust for financial support, and our colleagues and friends in Latin America for their generous collaboration over many years. James Patterson kindly provided Figure 1. We especially thank Iain Fram and Russell Stothard for early work on genetic hybridization

References

1. Miles MA. New world Trypanosomiasis. In: Microbiology and Microbial Infections. London: Topley and Wilson, 1997.
2. Gaunt M, Miles MA. The ecotopes and evolution of triatomine bugs (Triatominae) and their associated trypanosomes. Memórias do Instituto Oswaldo Cruz 2000; 95:5557-5565.
3. Miles MA, Toye, PJ, Oswald SC et al. The identification by isoenzyme patterns of two distinct strain-groups of Trypanosoma cruzi, circulating independently in a rural area of Brazil. Trans R Soc Trop Med Hyg 1977; 71:217-225.
4. Miles MA. Transmission cycles and the heterogeneity of Trypanosoma cruzi. In: Lumsden WHR, Evans DA, eds. Biology of the Kinetoplastida. Vol. 2. 1979:117-196.
5. Miles MA, Souza A, Povoa MM et al. Isozymic heterogeneity of Trypanosoma cruzi in the first autochthonous patients with Chagas disease in Amazonian Brazil. Nature 1978; 272:819-821.
6. Miles MA, Cedillos RA, Povoa, MM et al. Do radically dissimilar Trypanosoma cruzi strains (zymodemes) cause Venezuelan and Brazilian forms of Chagas disease? Lancet 1981; 20:1338-1340.
7. Povoa MM, De Souza AA, Naiff RD et al. Chagas disease in the Amazon Basin IV. Host records of Trypanosoma cruzi zymodemes in the States of Amazonas and Rondonia, Brazil. Ann Trop Med Parasitol 1984; 78:479-487.
8. Tibayrenc M, Miles MA. A genetic comparison between Brazilian and Bolivian zymodemes of Trypanosoma cruzi. Trans R Soc Trop Med Hyg 1983; 77:76-83.
9. Miles MA, Apt W, Widmer G et al. Isozyme heterogeneity and numerical taxonomy of Trypanosoma cruzi stocks from Chile. Trans R Soc Trop Med Hyg 1984; 78:526-535.
10. Chapman MD, Caffrey A, Swallow DM et al. Enzyme subunit numbers in Trypanosoma cruzi zymodemes. Ann Trop Med Parasitol 1984; 78:541-542.

11. Chapman MD, Baggaley RC, Godfrey-Fausset PF et al. Trypanosoma cruzi from the Paraguayan Chaco: Isoenzyme profiles of strains isolated at Makthlawaiya. J Protozool 1984; 31:482-486.
12. Brisse S, Barnabe C, Tibayrenc M. Identification of six Trypanosoma cruzi phylogenetic lineages by random amplified polymorphic DNA and multilocus enzyme electrophoresis. Int J Parasitol 2000; 30:35-44.
13. Morel C, Chiari E, Camargo EP et al. Strains and clones of Trypanosoma cruzi can be characterised by pattern of restriction endonuclease products of kinetoplast DNA minicircles. Proc Nat Acad Sci USA 1980; 77:6810-6814.
14. Tibayrenc M, Neubauer K, Barnabe C et al. Genetic characterization of six parasitic protoza: Parity between random-primer DNA typing and multilocus enzyme electrophoresis. Proc Nat Acad Sci USA 1993; 90:1335-1339.
15. Zingales B, Souto RP, Mangia RH et al. Molecular epidemiology of American trypanosomiasis in Brazil based on dimorphisms of rRNA and mini-exon gene sequences. Int J Parasitol 1998; 28:105-112.
16. Oliveira RP, Broude NE, Macedo AM et al. Probing the genetic population structure of Trypanosoma cruzi with polymorphic microsatellites. Proc Nat Acad Sci USA 1998; 95:3776-3780.
17. Carreno H, Rojas C, Aguilera X et al. Schizodeme analyses of Trypanosoma cruzi zymodemes from Chile. Exp Parasitol 1987; 64:252-260.
18. Fernandes O, Souto RP, Castro JA et al. Brazilian isolates of Trypanosoma cruzi from humans and triatomines classified into two lineages using mini-exon and ribosomal RNA sequences. Am J Trop Med Hyg 1998; 58:807-811.
19. Fernandes O, Sturm NR, Derre R et al. The mini-exon gene: A genetic marker for zymodeme III of Trypanosoma cruzi. Molecular and Biochemical Parasitology 1998; 95:129-133.
20. Mendonca MBA, Nehme NS, Santos SS et al. Two main clusters within Trypanosoma cruzi zymodeme 3 are defined by distinct regions of the ribosomal RNA cistron. Parasitology 2002; 124(Pt 2):177-84.
21. Clark CG, Pung OJ. Host specificity of ribosomal DNA variation in sylvatic Trypanosoma cruzi from North America. Mol Bioch Parasitol 1994; 66:174-179.
22. Anon. Recommendations from a Satellite Meeting Memórias do Instituto Oswaldo Cruz 1999; 94(Suppl 1):429-432.
23. Gibson WC, Stevens J. Genetic exchnge in the Trypanosomatidae. Adv Parasitol 1999; 43:1-46.
24. Tibayrenc M, Ward P, Moya A et al. Natural populations of Trypanosoma cruzi, the agent of Chagas disease, have a complex multiclonal structure. Proc Nat Acad Sci USA 1986; 83:115-119.
25. Spratt BG, Maiden MC. Bacterial population genetics, evolution and epidemiology. Philos Trans R Soc Lond B Biol Sci 1999; 354:701-710.
26. Widmer G, Dvorak JA, Miles MA. Temperature modulation of growth rates and glucosephosphate isomerase isozyme activity in Trypanosoma cruzi. Mol Biochem Parasitol 1987; 23:55-62.
27. Machado C, Ayala FJ. Nucleotide sequences provide evidence among distantly related lineages of Trypanosoma cruzi. Proc Nat Acad Sci USA 2001; 98:7396-7401.
28. Miles MA. Ploidy, heterozygosity and antigenic expression of South American trypanosomes. Parassitologia 1985; 27:87-104.
29. Carrasco HJ, Frame IA, Valente SA et al. Genetic exchange as a possible source of genomic diversity in sylvatic populations of Trypanosoma cruzi. Am J Trop Med Hyg 1996; 54:418-424.
30. Tait A. Nature 1980; 287:536-538.
31. MacLeod A, Tweedie A, Welburn SC et al. Minisatellite marker analysis of Trypanosoma brucei. Reconciliation of clonal, panmictic, and epidemic population genetic structures. Proc Nat Acad Sci USA 2000; 24:13442-13447.
32. Stothard JR, Frame IA, Miles MA. Genetic diversity and genetic exchange in Trypanosoma cruzi: dual drug-resistant "progeny" from episomal transformants. Memórias do Instituto Oswaldo Cruz 1999; 94(Suppl 1):189-193.
33. McDaniel JP, Dvorak JA. Identification, isolation and characterization of naturally occurring Trypanosoma cruzi variants. Mol Biochem Parasitol 1993; 57:213-222.
34. Dvorak J, Hall T, Crane MSTJ et al. Trypanosoma cruzi: Flow Cytometric Analysis. I. Analysis of total DNA/organism by means of mithramycin-induced fluoresence. J Protozool 1982; 29:430-437.

35. Mondragon A, Wilkinson SR, Taylor MC et al. Optimization of conditions for growth of wild-type and genetically transformed Trypanosoma cruzi on agarose plates. Parasitology 1999; 118:461-467.
36. Fernandes O, Mangia RH, Lisboa CV et al. The complexity of the sylvatic cycle of Trypanosoma cruzi in Rio de Janeiro state (Brazil) revealed by the non-transcribed spacer of the mini-exon gene. Parasitology 1999; 118:161-166.
37. Briones MRS, Souto RP, Stolf BS et al. The evolution of two Trypanosoma cruzi subgroups inferred from rRna genes can be correlated with the interchange of American mammalian faunas in the Cenozoic and has implications to pathogenicity and host specificity. Mol Biochem Parasitol 1999; 104:219-232.
38. Dvorak JA. The natural heterogeneity of Trypanosoma cruzi: biological and medical implications. J Cell Biochem 1984; 24:357-371.
39. Luquetti AO, Miles MA, Rassi A et al. Trypanosoma cruzi: zymodemes associated with acute and chronic Chagas disease in central Brazil. Trans R Soc Trop Med Hyg 1986; 80:462-470.
40. Vago AR, Andrade LO, Leite AA et al. Genetic characterization of Trypanosoma cruzi directly from tissues of patient chronic Chagas disease: differential distribution of genetic types into diverse. Am J Pathol 2000; 156:1805-1809.
41. Robello C, Gamarro F, Castanys S et al. Evolutionary relationships in Trypanosoma cruzi: molecular phylogenetics supports the existence of a new major lineage of strains. Gene 2000; 246:331-338.
42. Andersson B, Aslund L, Tammi M et al. Complete sequence of a 93.4 kb contig from chromosome 3 of Trypanosoma cruzi containing a strand-switch region. Genome Res 1998; 8:811-815.

CHAPTER 2

Distinct Mechanisms Operate to Control Stage-Specific and Cell-Cycle Dependent Gene Expression in *Trypanosoma cruzi*

Maria Carolina Q. Barbosa Elias, Rafael Marques Porto, Marcella Faria and Sergio Schenkman

Abstract

Proliferation and differentiation are key events for the establishment of infection by *Trypanosoma cruzi* and consequently, for the pathogenesis of Chagas disease. Therefore the understanding of these processes at the molecular level is important for the design of new prophylactic and therapeutic strategies to combat Chagas disease. Very little is known about the mechanisms that control these processes, which involve the transformation of proliferative and noninfective to nonproliferative and infective forms. In this chapter, we will initially summarize the morphological differences between the parasite stages, highlighting some of the biochemical changes at the cell surface. Then, we will describe some of the known mechanisms involved in the control of differential gene expression. Finally, we will discuss the changes that occur to the nuclear and chromatin structure when the proliferating stages differentiate into infective stages. Based on the recent findings, we will propose two levels for the control of gene expression in *T. cruzi*. The first regulates gene expression by controlling mRNA stability. In this case, environmental signals may induce, activate, or inactivate factors, such as RNA binding proteins, which regulate the steady-state levels of each individual mRNA. The second level of control involves the triggering by environmental signals of the entry and exit of the cell cycle. These events are associated with structural modifications of the nucleus and the kinetoplast affecting the replication machinery and, nonspecifically, transcription.

Introduction

T. cruzi cycles between proliferative and infective stages. Accumulated knowledge clearly points out that the transitions between these stages are brought about by changes in the environment. The parasite proliferates inside the insect vector gut (as epimastigote forms), or in the mammalian cell cytoplasm (as amastigote forms). These environments can be considered "nutrient rich". When nutrients become scarce, in the hindgut of the insect, or when the mammalian cell cytoplasm is full of parasites, proliferation stops and the parasites differentiate into infective trypomastigote forms. The infective, or differentiated forms are adapted in preparation for the next stage of the life cycle when they regain access to a nutrient rich environment. These forms of the parasite have restricted metabolic functions until they re-enter the cell

Molecular Mechanisms in the Pathogenesis of Chagas Disease, edited by John M. Kelly.

cycle. Cell-derived trypomastigotes circulate in the bloodstream of infected individuals. They then invade a new mammalian cell and transform into the replicative amastigote forms or are taken up by the insect vector, transforming into epimastigotes. Insect-derived trypomastigotes (metacyclic-trypomastigotes) must infect a mammalian cell to continue the parasite life cycle. During these transitions several morphological changes are observed, with different patterns of gene expression. However, very little is known about the mechanisms that underlie the control of this stage-specific gene expression or the molecular events that induce the morphological changes.

Morphological Changes during Differentiation

The life cycle of *T. cruzi* involves replicative and infective forms, which present morphological and biochemical differences.[1] The most evident morphological structures are characteristic of each stage and include the cell shape, the position of the kinetoplast in relation to the nucleus, and the region where the flagellum emerges from the flagellar pocket. In all stages the basal body, from which the flagellum originates, is located close to the kinetoplast (Figure 1).

In replicating epimastigotes and amastigotes forms, the kinetoplast is located anterior to the nucleus, and it contains filamentous material arranged in a packed row of fibers oriented parallel to the longitudinal axis of the parasite. This structure appears as a slightly concave disk of 1 μm in length, and 0.1 μm in depth. Epimastigotes are spindle-shaped organisms, 20-40 μm long and amastigotes are rounded forms 3-5 μm in diameter. The infective and nonreplicating trypomastigote forms have a length of about 25 μm and a diameter of about 2 μm. In contrast to replicating stages, the kinetoplast of trypomastigotes is located posterior to the nucleus, it is round and the filaments are in a more dispersed state.

The flagellum size is not related to the cell cycle and also differs among the parasite stages. Amastigotes have a short flagellum, 1 μm in length, located inside the flagellar pocket. It grows up to about 20 μm during the intracellular differentiation to trypomastigotes. Proliferating epimastigote forms always have a long flagellum. The plasma membrane composed of proteins, lipids and carbohydrates which form the glycocalix, also changes during the life cycle of the parasite. Cytochemical and electron microscopy studies show that in trypomastigotes the glycocalix is about 15 nm thick while in amastigotes and epimastigotes the thickness is about 5 nm. We have found that this glycocalix is formed by a layer of mucin-like molecules. These glycoproteins are sialylated by a surface enzyme with *trans*-sialidase activity (see below) and act to protect the parasite from osmotic changes, proteases, agglutinins, complement, and oxidants.[2]

The nucleus also undergoes large structural changes during differentiation. In both replicative forms the nucleus is round, contains a large nucleolus and the heterochromatin is concentrated at nuclear periphery. In contrast, in the nonreplicative infective forms the nucleus is elongated, the nucleolus is not evident and the heterochromatin is dispersed in the nuclear space[3] (Figure 1). This nuclear reorganization reflects a general decrease in the transcription of all genes of the parasite. The transcription of the ribosomal genes is diminished and the nucleolus disassembles.[3] RNA polymerase II transcription is also reduced, and this may explain the large increase in the heterochromatin seen in trypomastigote forms.[3]

Differential Expression of Surface Glycoproteins

Due to the fact that the cell surface is the interface between the parasite and the environment, many groups have been working to characterize the structure and function of surface molecules in the different developmental stages of *T. cruzi*. These studies indicate that this parasite differentially expresses several surface glycoproteins.[4,5] In the following sections we will provide a general description of the major surface glycoproteins of the *T. cruzi* and discuss how they are differentially expressed.

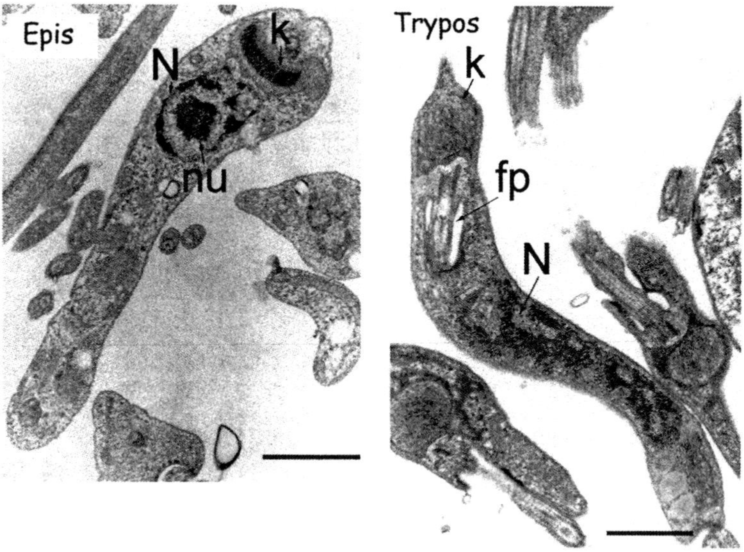

Figure 1. Electron microscope images of *T. cruzi* show differences in the nucleus of epimastigotes and cell-derived trypomastigote forms. Epimastigote nuclei (N) contain a large and central nucleolus (nu). The heterochromatin, seen as the electron dense structures close to the nuclear envelope represents a small portion of the nuclear space in epimastigotes. In about 50% of epimastigotes the heterochromatin is located more centrally, close to the nucleolus. In contrast, the nucleolus is not distinguishable from the heterochromatin that occupies a large portion of the nuclear space. In the figure the typically spherical kinetoplast (K) is visible in the trypomastigote form. In epimastigotes the kinetoplast has a cylindrical form and is attached to the flagellar pocket (fp). The bar = 1 μm.

Mucin-Like Glycoproteins

The most abundant glycoproteins of *T. cruzi* are threonine rich mucin-like molecules containing sialylated *O*-linked oligosaccharides with terminal α or β-galactose.[5,6] In contrast to mammalian mucins, the oligosaccharides are *O*-linked through a N-acetyl-glucosamine instead of a typical N-acetyl-galactosamine. We estimate that there are 10^6 to 10^7 mucin-like glycoproteins per parasite, depending on the stage. These glycoproteins are highly hydrophilic and are anchored on the surface by glycosyl-phosphatidylinositol. The presence of sialic acid is responsible for the negatively charged coat surrounding the parasite surface and this has primarily a structural role.[7] The coat of mucin molecules protects the parasite from antibody-mediated lysis, proteases and oxidative agents. The sialylation must also affect the interaction of the parasite with their host, particularly during cell adhesion, invasion, escape from the parasitophagous vacuole and migration through the infected host.[8]

The type of mucin expressed in the different stages of the parasite varies. In insect forms (epimastigotes and metacyclic trypomastigotes), the mucins are short glycoproteins with a simpler oligosaccharide structure.[7,9] These short mucins are highly resistant to proteases and act to protect the parasite from the action of digestive enzymes in the insect gut. In trypomastigotes derived from mammalian cells, the mucins are much larger, with more complex type *O*-oligosaccharides,[10,11] which could explain the presence of a thicker coat in this stage.[7] The lipid anchor of mucins also varies. While in the epimastigotes they are formed by alkyl-acyl phospholipids, they contain ceramide in metacyclic-trypomastigotes[12] and unsaturated fatty acids in cell-derived trypomastigotes.[11] The protein core of the mucins is also developmentally regulated. It is encoded by a very large family of genes (>500 copies). The products of these genes have highly conserved N- and C-termini and variable internal domains, composed of either conserved threonine-rich repeats or threonine-rich random sequence.[13,14] Both types of gene are expressed in trypomastigotes, but only the mucins containing the repeats are *O*-glycosylated.[15] A new group of threonine-rich proteins has recently been found, which correspond to the mucins preferentially expressed in epimastigotes.[16] Therefore, the differential expression of mucins, their post-translational modifications and the structure of the lipid anchor are controlled at each stage by mechanisms that are independent of the replication status of the parasite.

Sialidase Super-Family of Glycoproteins

A large class of *T. cruzi* surface glycoproteins in the size range 80 to 200 kDa are characterized by conserved motifs, the SXDXGXTW (the so-called Asp-box) and the VTVXNVXLYNR sequences.[17,18] Asp-boxes are common in nonviral sialidases, forming loops connecting some β-sheets that surround the barrel-like structures of these proteins.[19] The second motif is unique and highly conserved in *T. cruzi*. It forms a lectin-binding domain, suggesting that it might have an important role in parasite adhesion.[20]

Glycoproteins of about 85 kDa represent a large sub-family of the sialidase super-family. These are highly variable glycoproteins encoded by about 800 different genes differentially expressed in the various stages of the *T. cruzi* life cycle. Some members of this gene family are found to encode receptors for parasite attachment during the invasion of mammalian cells.[4] The ligands for this group of proteins are extracellular matrix components such as fibronectin,[21] laminin,[22] collagen,[23] cytokeratins,[20] and carbohydrates present on the host cell surface.[24] Some members of this large group of glycoproteins are expressed specifically in mammalian cell-derived trypomastigotes, others in insect-derived trypomastigotes,[25] and yet others in amastigotes.[26] This differential expression could reflect the different types of interactions that these parasite forms have with their respective host cells.

A smaller group of these glycoproteins are enzymes with trans-sialidase (TS) activity encoded by about 80 genes.[18] The TS transfers sialic acid from host glycoconjugates to the mucin-like glycoproteins of the parasite surface. The TS sequence contains one VTVXNVXLYNR and five SXDXGXTW motifs. Their position can be clearly localized in the X-ray deduced structure of *Trypanosoma rangeli* sialidase.[27] Distinct from other groups of glycoproteins, the TS contains an amino acid repeat that promotes enzyme oligomerization at its C-terminus.[28] As the enzyme is anchored to the parasite surface by a glycosyl-phosphatidylinositol,[29] the C-terminal repeats may project the enzyme far from the surface, helping in the sialylation/desialylation process either of the parasite, or the host surface glycoproteins. The large mucins expressed in trypomastigote stages would be particularly suitable substrates.

Most of the TS activity is detected in trypomastigotes released from the mammalian cells although expression of these molecules begins prior to lysis of the host cell.[30] The enzyme is released by the parasite, accumulates initially inside the infected mammalian cell cytoplasm, and is later released into the body fluids of the infected host. Insect forms of *T. cruzi* also express TS activity, although at much lower level (500 times lower).[31] TS activity increases very little in insect-trypomastigotes. The enzyme expressed in these stages lacks the C-terminal repeats and is not released from the parasite surface.

There is evidence showing that TS is involved in cell adhesion and invasion,[32,33] that it functions in parasite escape from the cell vacuole,[34] and that it acts as an important virulence factor of the parasite.[35] The fact that TS is already produced while the parasite is inside the mammalian cell also suggests that it could be involved in parasite escape.[8] Sialylation of the mucins on the parasite surface, in addition to increasing parasite resistance against proteases and lytic factors, can certainly affect parasite migration and dispersion in the host. Moreover, the fact that mammalian hosts produce antibodies that inhibit TS activity and parasite sialylation in the initial stages of the infection,[36] could be relevant for pathogenesis. Nevertheless, direct evidence that the TS has a critical role in the parasite life cycle, establishment of infection, or pathogenesis remains to be unequivocally demonstrated.

As in the case of the mucins, the differential expression of the sialidase super-family is a consequence of adaptive responses in each parasite form and seems to be unrelated to the parasite cell cycle.

Control of Gene Expression

In the following section we will describe some regulatory features of gene expression in *T. cruzi,* highlighting those involved in the expression of the surface proteins.

Transcription Control

Classical promoters for RNA polymerase II and transcription initiation and termination sites have not been identified in trypanosomes. Transcription seems to occur on very long DNA segments, including several coding regions and intergenic spacers.[37,38] There is much evidence showing that transcription is constitutive and that gene expression is controlled post-transcriptionally (see below). mRNA precursors are processed by *trans*-splicing and poly-adenylation. The *trans*-splicing reaction is similar to *cis*-splicing, but uses an exon from a separate RNA that is called the splice leader. The splice leader RNA undergoes the capping reaction and is a 5'-donor in the *trans*-splicing reaction. Poly-adenylation occurs at the 3' end of the precursor mRNA. Both processes are coupled and generate mature mRNA molecules that are exported to the cytoplasm.[39] It is unclear whether the splice-leader is transcribed by an RNA polymerase II-like enzyme,[40] or by an enzyme similar to the type III RNA polymerase.[41]

The RNA polymerase I, which transcribes the ribosomal genes, and some telomere-associated genes in African trypanosomes, can be used artificially to generate mRNA in *T. cruzi.* In these cases the transcripts are normally processed by *trans*-splicing and poly-adenylation,[42-45] suggesting that transcription is not necessarily linked to splicing and poly-adenylation, as it is in most eukaryotes.[46] A possible explanation for this uncoupling is that the C-terminal domain (CTD) of the large subunit of the RNA polymerase II of higher eukaryotes contains heptapeptide repeats, which are absent in trypanosomes. The CTD is a key element in the coupling between transcription, splicing, and poly-adenylation.[46-48] We have sequenced the largest subunit of RNA polymerase II of *T. cruzi* (GenBank AF372503) and shown that it lacks the heptapeptide repeats, as in other trypanosomes.[49-51] However, when the CTDs of trypanosome RNA polymerases II are aligned, several conserved motifs become evident,[51A] suggesting that the CTD domain of these parasites might have a role during the control of transcription initiation. As the enzyme is phosphorylated,[52] these conserved sequences could be substrates for stage-specific and regulatory kinases. On the other hand, the lack of repeats, which are known to provide the binding sites for the splicing, poly-adenylation and termination factors might explain the polycystronic nature of precursor mRNA in trypanosomes. The absence of repeats in the RNA polymerase II of higher eukaryotes leads to enzyme run-through of the poly-adenylation and termination sites, a situation that normally occurs in trypanosomes.

As mentioned above, data have suggested that transcription occurs constitutively in *T. cruzi.* By using lysolecithin permeable cells, or isolated nuclei to incorporate RNA precursors, we have demonstrated that transcription in *T. cruzi* is constitutive for all genes,[53] including those that are not processed into mRNA.[3] For example satellite genes, 195 bp repeats, which correspond to about 10% of *T. cruzi* DNA are constitutively transcribed. Nevertheless, transcription levels are proportionally reduced for this and other genes present in a large copy number.[3] There is no current explanation for these findings, but it is possible that chromatin structure and the nuclear environment modulate transcription of these highly repetitive genes, as found in most eukaryotes. Highly repetitive genes can form a more compact chromatin structure less accessible to the transcription machinery; consistent with the fact that heterochromatin is clearly visible in the *T. cruzi* nucleus (see Figure 1). Silencing in repetitive genes might be also attributed to chromatin modifications, such as β-D-glucosyl-hydroxymethyluracil found in the DNA of kinetoplastid protozoans,[54] methylation or histone modifications,[55] described in other eukaryotes. Alternatively, transcription of genes present in low copy numbers could be up-regulated by an unknown mechanism.

Post-Transcriptional Control by RNA Stability

The mechanisms that regulate mRNA levels in *T. cruzi* are mainly related to transcript stability. The stability may be controlled just after transcription, during RNA splicing, transport to the cytoplasm, or in the cytoplasm at translational initiation. As in other kinetoplastids, the presence of specific sequences in the 3' untranslated region (UTR) of the mRNAs, have been found to control transcript stability in most of the genes that have so far been studied.[16,56-59] Experiments have shown that introducing specific 3' -UTR sequences derived from amastin, tuzin, mucin, and gp85 genes, amongst others, can modulate the expression of reporter genes in *T. cruzi.*

Studies from our own, and other laboratories have shown that stage-specific expression in *T. cruzi* is related primarily to changes in mRNA stability. As an example, amastin mRNA levels decrease, while the TS mRNA levels increase when intracellular amastigotes differentiate into trypomastigotes inside the mammalian cell.[53] The TS mRNA remains stable until the parasite is released from infected cells, when it then starts to decay. In contrast, the *gp85* mRNA becomes stabilized only after the parasites are released from the infected cell. The fact that cycloheximide, a protein synthesis inhibitor, stabilizes these mRNAs indicates that stability occurs in the cytoplasm. Moreover, cycloheximide increases the stability of TS and gp85 mRNAs only before their steady-state levels are maximal. It does not prevent the decay, suggesting that stability of each mRNA is controlled differently, and the mechanisms acting on the stabilization are distinct from the ones that control the decay.

Analysis of the mucin mRNA stability has revealed that stage-specific RNA-binding proteins differentially recognize AU-rich elements (ARE) in the 3' UTR of mRNA of mucins expressed in epimastigotes and trypomastigotes.[60] In addition these studies revealed the presence of a *cis* acting G-rich element in the 3' UTR which is responsible for general mRNA decay. A differentially expressed RNA binding protein seems to be involved in protection of the mRNA from degradation.

More detailed studies on the specific mechanisms that control the stability of each mRNA species are required. It will also be important to identify and dissect the signaling pathways that operate during differentiation and to discover how they interact with the regulatory elements/ proteins that govern mRNA stability. The nature of these signaling pathways are unknown, and might be related to starvation through G-protein,[61,62] Ca^{2+}, phosphoinositide or cAMP dependent mechanisms.[63] Factors also known to affect differentiation are protein and nucleic acid metabolites such as amino acids, peptides, purines, pyrimidines, and polyamines.

Nuclear Changes and the Cell Cycle

As mentioned above, a remarkable nuclear and chromatin reorganization is observed when reproductive forms of *T. cruzi* transform into the infective forms (Figure 1). These modifications correlate with a general down-modulation of the transcription machinery and probably reflect that the reproductive forms of the parasite require large amounts of newly synthesized protein to grow and replicate. However, these changes in the nuclear structure do not seem to correlate with the differential gene expression in the various stages of *T. cruzi*, such as those described above. The nuclear morphology is similar in the metacyclic-trypomastigotes and cell-derived trypomastigotes, which present a different pattern of expressed genes. Also, both proliferating forms (epimastigotes and intracellular amastigotes) have a quite similar nuclear structure and present a rather different gene expression pattern. Therefore, the nuclear organization seems to be more related to whether the parasites are replicating, engaged in cell cycling, or whether they are nonreplicating and in the differentiated trypomastigote stages.

To compare the nuclear structure in these different parasite stages, we have utilized fluorescent *in situ* hybridization (FISH) using satellite DNA as a probe. In *T. cruzi* the satellite sequences are formed by 195 base pairs present in about 10^5 copies, corresponding to about 10% of the parasite genome.[64] FISH analysis shows that the highly repetitive satellite DNA is present in dots homogeneously dispersed in the nucleus of all trypomastigotes, while in half of actively growing epimastigotes, the satellites are dots distributed in the periphery of the nucleus (Figure 2). These two types of satellite DNA distribution in the epimastigotes correlate with the cell division cycle of the parasite. The pattern in trypomastigotes seems to correspond to those of the epimastigote population that are in the G1 phase of the cell cycle and show a corresponding dispersed pattern of satellite signal.

As observed by electron microscopy the nucleus of epimastigotes contains 10 small electron-dense plaques.[65] These plaques change position in the nucleus, migrating to polar regions during mitosis. The nature of these plaques is not known, but it has been proposed that they could correspond to kinetochore-like structures, playing an important role in the process of separation of the nuclear material into the two daughter cells. It is possible (although not yet fully demonstrated) that the *T. cruzi* kinetochore could be composed of satellite DNA, as found in higher eukaryotes.[66] By studying satellite DNA organization in the *T. cruzi* genome, we found that these sequences are present in 10-12 chromosomes, forming long tandem repeats. These repeats may form large supra-chromatin or heterochromatic structures, characteristic of centromeric repeated DNA in eukaryotes. This idea is compatible with the fact that the satellite DNA repeat is highly conserved (more than 90%), and a preferential segment of the 195 base pair repeat is protected in the nucleosomes from *Micrococcus* nuclease digestion (unpublished results). In addition, it is conceivable that the satellite sequences could be localized in the dense plaques, since (i) there are 10 plaques and there are 10-12 chromosomes containing satellites, (ii) the supra-chromatin structure formed by satellite sequences can be visualized as more dense regions by electronic microscopy, and (iii) we also identified 10 to 12 spots by FISH that migrate to polar regions before the onset of mitosis (Figure 2).

The changes in nuclear structure observed in replicating parasites as a result of the alterations in the position of heterochromatin and dense plaques suggest that the chromatic material moves during the cell cycle of replicating forms. This movement is also evidenced by our satellite DNA FISH analysis (Figure 2). In G1 phase, the satellite sequences are dispersed in the entire nuclear space as found in trypomastigotes. When the cells enter into S-phase, the satellite DNA moves to the nuclear periphery and remains there until the end of mitosis when the labeling disperses in the two daughter cells. Incorporation of bromo-deoxy-uridine followed by immunofluorescence analysis with anti-bromo-deoxy-uridine antibodies revealed that most of the replication sites were in the nuclear periphery. After the end of the S-phase the bromo-deoxy-uridine labeling reappeared in the nuclear interior, suggesting that satellite DNA had moved from the center to the nuclear periphery, possibly to allow chromosome replication.[67]

Therefore, the morphological changes in the nuclear structure appear to be a consequence of whether the parasite is engaged in the cell cycle. By inference these changes must be induced

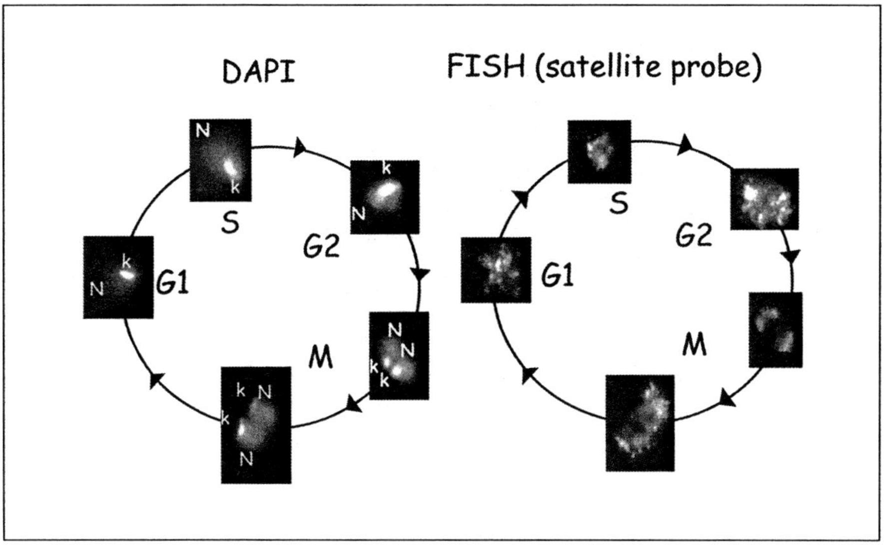

Figure 2. FISH analysis made with a satellite DNA probe shows different chromosomal distribution during the cell cycle and differentiation of *T. cruzi*. Parasites were attached to silane coated glass slides and processed for FISH using a digoxygenin-PCR labeled satellite fragment. Using anti-digoxigenin antibodies and a FITC-anti-IgG we detected the hybridization signals. The slides were also mounted with DAPI to visualize the DNA from the nucleus (N) and the kinetoplast (k), which allow us to distinguish the cell cycle phases: G1, S, G2 and M.

and controlled, either directly or indirectly, by the signals that regulate the cell cycle. It is also implicit that these signals affect the changes in the kinetoplast organization. In contrast, differences in the flagellar size, microtubule organization, and cell surface composition would appear to be controlled by other signaling systems, as described above.

Chromatin Modifications during the Cell Cycle

Variations in the nuclear organization are normally associated with modifications in chromatin structure,[68] usually generated by histone modifications.[55,69] In *T. cruzi* the nuclear DNA is organized in a typical nucleosomal pattern. The disk-like nucleosomal core is formed by two of each of the four core histones, H2a, H2b, H3 and H4 as in higher eukaryotes. The core histones were initially identified as a, b, c and d in *T. brucei*, and based on gel electrophoresis and partial amino acid sequences they were considered to be equivalents of histones H3, H2A, H2B and H4, respectively.[70] Although histone H3 and especially histone H4 are known to be amongst the most conserved proteins in higher eukaryotes, in *T. cruzi* and other kinetoplastids these histones differ by 50% in amino acid sequence when compared to their human counterparts. Histones H2A and H2B show about 45-48% identity to the same histones of higher eukaryotes.[71]

Histone H1, which is required to achieve two complete turns of the DNA around the nucleosome core and is thought to facilitate packing of the chromatin in its most condensed form, is the most divergent *T. cruzi* histone. It is similar to the C-terminal region of the histone H1 of higher eukaryotes, and lacks the central globular and the N-terminal domain.[72-74] The absence of the globular domains may be responsible for the fact that the *T. cruzi* chromatin never condenses to the point of 30 nm fibers during mitosis.

Very little is known about how the histones change during the cell cycle and differentiation of *T. cruzi,* or whether there are chemical modifications similar to those commonly found in the histones of other eukaryotes.[68] *T. cruzi* core histones, and most of histone H1 are synthesized concomitantly with DNA replication, as in higher eukaryotes, but some histone H1 is constitutively expressed.[71] Histone genes are also preferentially expressed in proliferating stages at the S-phase of the cell cycle in *Leishmania infantum*[75] and in *T. brucei*.[68,76]

We have found that there are important variations in the histone composition during differentiation and the cell cycle of *T. cruzi.* This was demonstrated by separating histones by Triton, acid, urea polyacrylamide gel electrophoresis (TAU-PAGE), which separates proteins according to their level of hydrophobicity. Two forms of histone H1 were observed in the TAU-PAGE, one phosphorylated and more abundant in G1 and nonproliferative trypomastigote forms, and the other present in proliferating forms.[51A]

In parallel to, and possibly as a consequence of changes in the histone repertoire, we have observed that chromatin is differentially sensitive to digestion by *Micrococcus* nuclease in proliferating versus nonproliferating forms of *T. cruzi.* The chromatin of trypomastigotes is approximately ten times more sensitive to this enzyme than the chromatin of epimastigotes. Differences in H1 phosphorylation could to a certain extent explain this finding, but it is likely that other factors are also involved.

Evidence suggests that protein phosphatases[77-79] and protein kinases[80,81] participate in the growth and differentiation control of *T. cruzi.* This supports the notion that phosphorylation events could be involved in the chromatin reorganization and possibly in the nuclear structure changes. *T. cruzi* cell division is regulated by Cdc related protein kinases[82] while differentiation might be related to cyclic AMP dependent signals.[83] Knowledge of chromatin modifications, as well as those occurring in the nuclear matrix and nuclear skeleton may therefore help to give a better understanding of the mechanisms that control entry and exit into the cell cycle in *T. cruzi.*

Conclusion

T. cruzi alternates continuously between cell proliferation and cell differentiation. Here, we have identified some morphological, physiological and biochemical changes that are associated with these transitions. As shown in Figure 3, the transitions between the proliferating and quiescent forms are characterized by structural reorganization of the nucleus, which is related to the events that control the activation or inactivation of the machineries involved in replication and transcription. The transcriptional regulation during these transitions is not specific for particular genes and reflects a general adaptive response to promote cellular growth through the synthesis of rRNA and the constitutive transcription of most chromosomal genes. In contrast, expression of stage-specific genes is mainly controlled at the post-transcriptional level, probably directed by environmental signals. These modifications allow the expression of a set of genes that facilitate adaptive responses to changing environmental conditions. The knowledge of how the environmental factors influence these two sets of control mechanisms will allow us to understand the critical steps in the biology of *T. cruzi,* and consequently the pathogenesis of Chagas disease.

Acknowledgments

Work in our laboratory is supported by Grants from FAPESP and CNPq (Brazil).

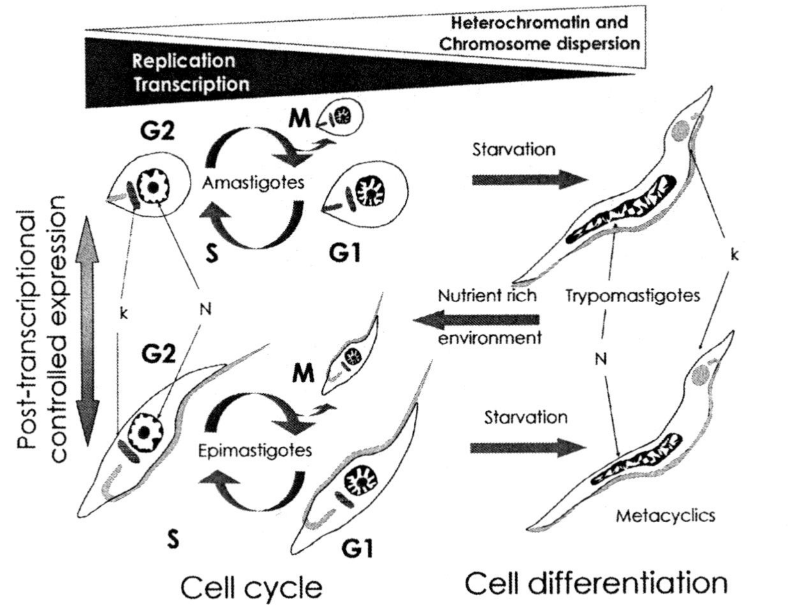

Figure 3. Cell cycle and differentiation of *T. cruzi*. The figure shows schematically the transitions between the four cell cycle stages, (G→S→G2→M) of epimastigotes and amastigotes forms. Note the round nucleus (N) with peripheral heterochromatin and an elongated kinetoplast (k) in G2 phase. Heterochromatin increases in G1 phase and in both trypomastigote forms. The differentiation of amastigotes and epimastigotes into trypomastigotes and metacyclic-trypomastigotes probably occurs at the G1 stage triggered by environmental signals originating most likely from a nutrient poor environment. A nutrient rich environment reintroduces the parasite into the cell cycle. The vertical double arrow indicates that other sets of environmental signals promote the post–transcriptional control of gene expression. Note the thick coat found specifically in the trypomastigote forms (see text for details). The arrow at the top indicates that the transcription and replication decrease when cells differentiate, while heterochromatin and chromosome dispersion is maximal in the differentiated forms.

References

1. de Souza W. Cell biology of Trypanosoma cruzi. Int Rev Cytol 1984; 86:97-283.
2. Chioccola VLP, Acosta-Serrano A, Almeida IC et al. Mucin-like molecules form a negatively charged coat that protects Trypanosoma cruzi trypomastigotes from killing by human anti-α-galactosyl antibodies. J Cell Sci 2000; 113(7):1299-1307.
3. Elias MC, Marques-Porto R, Freymuller E et al. Transcription rate modulation through the Trypanosoma cruzi life cycle occurs in parallel with changes in nuclear organisation. Mol Biochem Parasitol 2001; 112(1):79-90.
4. Alves MJ. Members of the Tc-85 protein family from Trypanosoma cruzi are adhesion proteins. Braz J Med Biol Res 1996; 29(7):831-833.
5. Frasch AC. Functional diversity in the trans-sialidase and mucin families in Trypanosoma cruzi. Parasitol Today 2000; 16(7):282-286.
6. Acosta-Serrano A, Almeida IC, Freitas-Junior LH et al. The mucin-like glycoprotein super-family of Trypanosoma cruzi: strucutre and biological roles. Mol Biochem Parasitol 2001; 114(2):143-150.

7. Pereira-Chioccola VL, Acosta-Serrano A, Correia DA et al. Mucin-like molecules form a negatively charged coat that protects Trypanosoma cruzi trypomastigotes from killing by human anti-α-galactosyl antibodies. J Cell Sci 2000; 113(Pt 7):1299-1307.
8. Pereira-Chioccola VL, Schenkman S. Biological role of Trypanosoma cruzi trans-sialidase. Biochem Soc Trans 1999; 27(4):516-518.
9. Previato JO, Jones C, Gonçalves LPB et al. O-glycosidically linked N-acetylglucosamine-bound oligosaccharides from glycoproteins of Trypanosoma cruzi. Biochem J 1994; 301(1):151-159.
10. Almeida IC, Ferguson MAJ, Schenkman S et al. Lytic anti-a-galactosyl antibodies from patients with chronic Chagas disease recognize novel O-linked oligosaccharides on mucin-like glycosylphosphatidylinositol-anchored glycoproteins of Trypanosoma cruzi. Biochem J 1994; 304(3):793-802.
11. Almeida IC, Camargo MM, Procopio DO et al. Highly-purified glycosylphosphatidyl inositols from Trypanosoma cruzi are potent pro-inflamatory agents. EMBO J 2000; 19(7):1476-1485.
12. Acosta-Serrano A, Schenkman S, Yoshida N et al. The lipid structure of the GPI-anchored mucin-like sialic acid acceptors of Trypanosoma cruzi changes during parasite differentiation from epimastigotes to infective metacyclic trypomastigote forms. J Biol Chem 1995; 270(45):27244-27253.
13. Freitas-Junior LHG, Briones MRS, Schenkman S. Two distinct groups of mucin-like genes are differentially expressed in the developmental stages of Trypanosoma cruzi. Mol Biochem Parasitol 1998; 93(1):101-114.
14. Di Noia JM, D'Orso I, Aslund L et al. The Trypanosoma cruzi mucin family is transcribed from hundreds of genes having hypervariable regions. J Biol Chem 1998; 273(18):10843-10850.
15. Pollevick GD, Di Noia JM, Salto ML et al. Trypanosoma cruzi surface mucins with exposed variant epitopes. J Biol Chem 2000; 275(36):27671-27680.
16. Di Noia JM, D'Orso I, Sanchez DO et al. AU-rich elements in the 3'-untranslated region of a new mucin-type gene family of Trypanosoma cruzi confers mRNA instability and modulates translation efficiency. J Biol Chem 2000; 275(14):10218-10227.
17. Cross GA, Takle GB. The surface trans-sialidase family of Trypanosoma cruzi. Annu Rev Microbiol 1993; 46(1):385-411.
18. Schenkman S, Eichinger D, Pereira MEA et al. Structural and functional properties of Trypanosoma cruzi trans-sialidase. Annu Rev Microbiol 1994; 48(1):499-523.
19. Vimr ER. Microbial sialidases: does bigger always mean better? Trends Microbiol 1994; 2(8):271-277.
20. Magdesian MH, Giordano R, Ulrich H et al. Infection by Trypanosoma cruzi: identification of a parasite ligand and its host-cell receptor. J Biol Chem 2001; 276(22):19382-19389.
21. Ouaissi MA, Afchain D, Capron A et al. Fibronectin receptors on Trypanosoma cruzi trypomastigotes and their biological function. Nature 1985; 308:380-382.
22. Giordano R, Fouts DL, Tewari D et al. Cloning of a surface membrane glycoprotein specific for the infective form of Trypanosoma cruzi having adhesive properties to laminin. J Biol Chem 1999; 274(6):3461-3468.
23. Velge P, Ouaissi MA, Cornette J et al. Identification and isolation of Trypanosoma cruzi trypomastigote collagen-binding proteins: possible role in cell-parasite interaction. Parasitol 1988; 97:255-268.
24. Kahn SJ, Wleklinski M, Ezekowitz RA et al. The major surface glycoprotein of Trypanosoma cruzi amastigotes are ligands of the human serum mannose-binding protein. Infect Immun 1996; 64(7):2649-2656.
25. Araya JE, Cano MI, Yoshida N et al. Cloning and characterization of a gene for the stage-specific 82-kDa surface antigen of metacyclic trypomastigoates of Trypanosoma cruzi. Mol Biochem Parasitol 1994; 65(1):161-169.
26. Low HP, Santos MA, Wizel B et al. Amastigote surface proteins of Trypanosoma cruzi are targets for CD8+ CTL. J Immunol 1998; 160(4):1817-1823.
27. Buschiazzo A, Tavares GA, Campetella O et al. Structural basis of sialyltransferase activity in trypanosomal sialidases. EMBO J 2000; 19(1):16-24.
28. Schenkman S, Chaves LB, Pontes de Carvalho L et al. A proteolytic fragment of Trypanosoma cruzi trans-sialidase lacking the carboxy-terminal domain is active, monomeric and generates antibodies that inhibit enzymatic activity. J Biol Chem 1994; 269(11):7970-7975.

29. Agusti R, Couto AS, Campetella OE et al. The trans-sialidase of Trypanosoma cruzi is anchored by two different lipids. Glycobiol 1997; 7(6):731-735.
30. Frevert U, Schenkman S, Nussenzweig V. Stage-specific expression and intracellular shedding of the cell surface trans-sialidase of Trypanosoma cruzi. Infect Immun 1992; 60:2349-2360.
31. Chaves LB, Briones MRS, Schenkman S. Trans-sialidase from Trypanosoma cruzi epimastigotes is expressed at the stationary phase and is different from the enzyme expressed in trypomastigotes. Mol Biochem Parasitol 1993; 61(1):97-106.
32. Schenkman S, Eichinger D. Trypanosoma cruzi trans-sialidase and cell invasion. Parasitol Today 1993; 9(6):218-222.
33. Yoshida N, Dorta ML, Ferreira AT et al. Removal of sialic acid from mucin-like surface molecules of Trypanosoma cruzi metacyclic trypomastigotes enhances parasite-host cell interaction. Mol Biochem Parasitol 1997; 84(1):57-67.
34. Hall BF, Webster P, Ma AK et al. Desialylation of lysosomal membrane glycoproteins by Trypanosoma cruzi: A role for the surface neuraminidase facilitating parasite entry into the host cell cytoplasm. J Exp Med 1992; 176(2):313-325.
35. Pereira MEA, Zhang K, Gong Y et al. Invasive phenotype of Trypanosoma cruzi restricted to a population expressing trans-sialidase. Infect Immun 1996; 64(9):3884-3892.
36. Pereira-Chioccola VL, Schenkman S, Kloetzel J. Sera from chronic Chagasic patients and animals infected with Trypanosoma cruzi inhibit trans-sialidase by recognizing its catalytic domain. Infect Immun 1994; 62(7):2973-2978.
37. Tschudi C, Ullu E. Trypanosomatid protozoa provide paradigms of eukaryotic biology. Infect Agents Dis 1994; 3(4):181-186.
38. Teixeira SM. Control of gene expression in Trypanosomatidae. Braz J Med Biol Res 1998; 31(12):1503-1516.
39. Davis RE. Spliced leader RNA trans-splicing in metazoa. Parasitol Today 1996; 12:1233-40.
40. Gilinger G, Bellofatto V. Trypanosome spliced leader RNA genes contain the first identified RNA polymerase II gene promoter in these organisms. Nucleic Acids Res 2001; 29(7):1556-1564.
41. Gunzl A, Ullu E, Dorner M et al. Transcription of the Trypanosoma brucei spliced leader RNA gene is dependent only on the presence of upstream regulatory elements. Mol Biochem Parasitol 1997; 85(1):67-76.
42. Floeter-Winter LM, Souto RP, Stolf BS et al. Trypanosoma cruzi: can activity of the rRNA gene promoter be used as a marker for speciation? Exp Parasitol 1997; 86(3):232-234.
43. Martinez-Calvillo S, Lopez I, Hernandez R. pRIBOTEX expression vector: a pTEX derivative for a rapid selection of Trypanosoma cruzi transfectants. Gene 1997; 199(1-2):71-76.
44. Ramirez MI, Yamauchi LM, de Freitas LH et al. The use of the green fluorescent protein to monitor and improve transfection in Trypanosoma cruzi. Mol Biochem Parasitol 2000; 111(1):235-240.
45. Vazquez MP, Levin MJ. Functional analysis of the intergenic regions of TcP2beta gene loci allowed the construction of an improved Trypanosoma cruzi expression vector. Gene 1999; 239(2):217-225.
46. Hirose Y, Manley JL. RNA polymerase II and the integration of nuclear events. Genes Dev 2000; 14(12):1415-1429.
47. Corden JL, Patturajan M. A CTD function linking transcription to splicing. Trends Biochem Sci 1997; 22(11):413-416.
48. Proudfoot N. Connecting transcription to messenger RNA processing. Trends Biochem Sci 2000; 25(6):290-293.
49. Evers R, Hammer A, Kock J et al. Trypanosoma brucei contains two RNA polymerase II largest subunit genes with an altered C-terminal domain. Cell 1989; 56(4):585-597.
50. Evers R, Hammer A, Cornelissen AW. Unusual C-terminal domain of the largest subunit of RNA polymerase II of Crithidia fasciculata. Nuc Acid Res 1989; 17(9):3403-3413.
51. Smith JL, Levin JR, Ingles CJ et al. In trypanosomes the homolog of the largest subunit of RNA polymerase II is encoded by two genes and has a highly unusual C-terminal domain structure. Cell 1989; 56(5):815-827.
51A. Marques Porto R, Amino R, Elias MCQ etal. Histone H1 is phosphorylated in non-replicating and infective forms of Trypanosoma cruzi. Mol Biochem Parasitol 2002; 119(2):265-71.

52. Chapman AB, Agabian N. Trypanosoma brucei RNA polymerase II is phosphorylated in the absence of carboxyl-terminal domain heptapeptide repeats. J Biol Chem 1994; 269(7):4754-4760.
53. Abuin G, Freitas-Junior LHG, Colli W et al. Expression of trans-sialidase and 85 kDa glycoprotein genes in Trypanosoma cruzi is differentially regulated at the post-transcriptional level by labile protein factors. J Biol Chem 1999; 274(19):13041-13047.
54. van Leeuwen F, Taylor MC, Mondragon A et al. beta-D-glucosyl-hydroxymethyluracil is a conserved DNA modification in kinetoplastid protozoans and is abundant in their telomeres. Proc Natl Acad Sci USA 1998; 95(5):2366-2371.
55. Cheung WL, Briggs SD, Allis CD. Acetylation and chromosomal functions. Curr Opin Cell Biol 2000; 12(3):326-333.
56. Nozaki T, Cross GA. Effects of 3' untranslated and intergenic regions on gene expression in Trypanosoma cruzi. Mol Biochem Parasitol 1995; 75(1):55-67.
57. Teixeira SM, Kirchhoff LV, Donelson JE. Post-transcriptional elements regulating expression of mRNAs from the amastin/Tuzin gene cluster of Trypanosoma cruzi. J Biol Chem 1995; 270(38):22586-22594.
58. Tomas AM, Kelly JM. Stage-regulated expression of cruzipain, the major cysteine protease of Trypanosoma cruzi is independent of the level of RNA. Mol Biochem Parasitol 1996; 76(1-2):91-103.
59. Maranon C, Puerta C, Alonso C et al. Control mechanisms of the H2A genes expression in Trypanosoma cruzi. Mol Biochem Parasitol 1998; 92(2):313-324.
60. D'Orso I, Frasch AC. Functionally different AU- and G-rich cis-elements confer developmentally regulated mRNA stability in Trypanosoma cruzi by interaction with specific RNA-binding proteins. J Biol Chem 2001; 276(19):15783-15793.
61. Coso OA, Díaz Añel A, Martinetto H et al. Characterization of a G_i-protein from Trypanosoma cruzi epimastigote membranes. Biochem J 1992; 287(2):443-446.
62. Oz HS, Huang H, Wittner M et al. Evidence for guanosine triphosphate—binding proteins in Trypanosoma cruzi. Am J Trop Med Hyg 1994; 50(5):620-631.
63. Parsons M, Ruben L. Pathways involved in environmental sensing in trypanosomatids. Parasitol Today 2000; 16(2):56-62.
64. Gonzalez A, Prediger E, Huecas ME et al. Minichromosomal repetitive DNA in Trypanosoma cruzi: its use in a high-sensitivity parasite detection assay. Proc Natl Acad Sci USA 1984; 81(11):3356-3360.
65. Solari AJ. Mitosis and genome partition in trypanosomes. Biocell 1995; 19(2):65-84.
66. Choo KH. Centromerization. Trends Cell Biol 2000; 10(5):182-188.
67. Elias MC, Faria M, Mortara RM et al. Chromosome movement during the cell cycle is determined by the position of replication sites in Trypanosoma. unpublish.
68. Belli SI. Chromatin remodelling during the life cycle of trypanosomatids. Int J Parasitol 2000; 30(6):679-687.
69. Aalfs JD, Kingston RE. What does 'chromatin remodeling' mean? Trends Biochem Sci 2000; 25(11):548-555.
70. Hecker H, Betschart B, Bender K et al. The chromatin of trypanosomes. Int J Parasitol 1994; 24(6):809-819.
71. Sabaj V, Aslund L, Pettersson U et al. Histone genes expression during the cell cycle in Trypanosoma cruzi. J Cell Biochem 2001; 80(4):617-624.
72. Aslund L, Carlsson L, Henriksson J et al. A gene family encoding heterogeneous histone H1 proteins in Trypanosoma cruzi. Mol Biochem Parasitol 1994; 65(2):317-330.
73. Toro GC, Galanti N. H1 histone and histone variants in Trypanosoma cruzi. Exp Cell Res 1988; 174(1):16-24.
74. Toro GC, Galanti N, Hellman U et al. Unambiguous identification of histone H1 in Trypanosoma cruzi. J Cell Biochem 1993; 52(4):431-439.
75. Soto M, Requena JM, Quijada L et al. Organization, transcription and regulation of the Leishmania infantum histone H3 genes. Biochem J 1996; 318(3):813-819.
76. Garcia-Salcedo JA, Gijon P, Pays E. Regulated transcription of the histone H2B genes of Trypanosoma brucei. Eur J Biochem 1999; 264(3):717-723.
77. Bakalara N, Seyfang A, Baltz T et al. Trypanosoma brucei and Trypanosoma cruzi: Life cycle-regulated protein tyrosine phosphatase activity. Exp Parasitol 1995; 81:302-312.

78. Orr GA, Werner C, Xu J et al. Identification of novel serine/threonine protein phosphatases in Trypanosoma cruzi: a potential role in control of cytokinesis and morphology. Infect Immun 2000; 68(3):1350-1358.
79. Grellier P, Blum J, Santana J et al. Involvement of calyculin A-sensitive phosphatase(s) in the differentiation of Trypanosoma cruzi trypomastigotes to amastigotes. Mol Biochem Parasitol 1999; 98(2):239-252.
80. Sadigursky M, Santos-Buch CA. A novel receptor mediated ATP transport system regulated by tyrosine and serine/threonine phosphokinases in Trypanosoma cruzi trypomastigotes. Recept Signal Transduct 1997; 7(1):29-43.
81. Ogueta SB, Macintosh GC, Tellez-Inon MT. Stage-specific substrate phosphorylation by a Ca2+/calmodulin-dependent protein kinase in Trypanosoma cruzi. J Eukaryot Microbiol 1998; 45(4):392-396.
82. Gomez EB, Santori MI, Laria S et al. Characterization of the Trypanosoma cruzi Cdc2p-related protein kinase 1 and identification of three novel associating cyclins. Mol Biochem Parasitol 2001; 113(1):97-108.
83. Naula C, Seebeck T. Cyclic AMP signaling in trypanosomatids. Parasitol Today 2000; 16(1):35-38.

CHAPTER 3

The *Trypanosoma cruzi* Mucin Coat:

Structure, Regulation of the Expression and Relevance in the Host-Parasite Relationship

Javier M. Di Noia, Ivan D'Orso and Alberto Carlos C. Frasch

Summary

The cell surface of *Trypanosoma cruzi*, the agent of the Chagas disease, is covered by a family of highly O-glycosylated mucin-like glycoproteins. This coat protects the parasite from the immune response of the host and is involved in the invasion of mammalian cells. Two major heterogeneous groups have been identified within the mucin family; the 35-50 kDa group which is expressed in the parasite stages associated with the insect vector, and the 60-200 kDa group which is expressed in the stages present in the vertebrate host. The carbohydrate moieties of the parasite mucins display some differences when compared with those expressed by mammalian cells. The first monosaccharide attached to the protein core is N-acetylglucosamine instead of N-acetylgalactosamine. Furthermore, sialic acid is attached to the oligosaccharide side chains by a unique *trans*-sialidase activity on the external surface of the parasite, and not in the Golgi apparatus as in most eukaryotic cells. An unexpectedly large number of parasite mucins have now been found after analysis of the genes encoding the core proteins. Two large gene families were identified. The TcSMUG gene family has about 70-80 members and encode the mucins that are expressed during the insect stage of the life-cycle. The second family, TcMUC, is composed of about 500-700 members and includes the genes that are expressed in the vertebrate host. Unlike the mucins expressed in the insect stages, those expressed in the vertebrate stages are characterized by the presence of a hypervariable N-terminal region. This is proposed to have a role in immunoevasion. The formidable task of regulating stage-specific gene expression of these large gene families is achieved, at least in part, through regulation of mRNA stability. Regulatory *cis*-acting sequences in mucin transcripts and the *trans*-acting protein factors that bind to these elements have now been identified. Thus, *T. cruzi* makes use of about 1% of its genome and has a complex post-transcriptional regulatory mechanism to generate the mucin coat required for its survival.

Mucins and Mucin-Like Molecules in Vertebrate Cells

The mucins are an evolutionary diverse family of glycoproteins that are characterized by a high level of O-glycosylation. The reducing ends of a large number of short oligosaccharides form a glycosydic bond with the hydroxyl groups of Thr and Ser residues present in the core protein, so that sugars represent 40-80% of the molecular mass.[1] The first described mucins

Molecular Mechanisms in the Pathogenesis of Chagas Disease, edited by John M. Kelly.

were secreted or transmembrane molecules from epithelial cells. These are normally very large molecules composed of a single polypeptide of approximately 400 to 5000 amino acids. O-linked oligosaccharides are clustered in the central region of the polypeptide which is made up of a variable number of tandem repeats, very rich in Ser, Thr and Pro residues, known as PTS regions.[2] These regions, which represent the main proportion of the polypeptide chain, are in most cases flanked by Cys-rich regions that are involved in the multimerization of the secreted mucins. This is the source of their gel forming properties.[1] The genes that encode epithelial cell mucins are very large; several have been identified in humans and other vertebrate species. The deduced products of the human *muc* genes contain repeated central domains, but the sequence of this region is not conserved between different proteins. This is also the case when mucins from different vertebrate species are compared.[2] Similarities between these proteins are restricted to the flanking Cys-rich regions. However, they can be grouped as a family based on their common overall structure. The expression of mucin genes and their level and type of glycosylation are regulated in a tissue specific manner. Post-transcriptional regulation is very frequent.[1]

A new category of mucins with a different developmental origin has been described in the last decade. It includes highly O-glycosylated membrane molecules that are present in leukocytes and endothelial cells that participate in lymphocyte trafficking.[3] Many of these molecules have now been identified. Since the presence of sialic acid at the non-reducing end is a common feature, they are also referred to as sialomucins.[4] In these molecules, the PTS regions in the core protein usually lack tandemly repeated sequences and there are similarities between molecules in different animal species. The sialomucins share the physico-chemical properties of epithelial mucins, but are low molecular weight molecules of 50 to 240 amino acids.[1] Sialomucins are monomeric or, at most, form dimers that are membrane associated or possess transmembrane domains. These glycoproteins are called mucin-like molecules to differentiate them from the epithelial mucins. The oligosaccharides of sialomucins act as ligands for lectins, participating in the adhesion mechanisms between blood cells and the endothelium.[4] Terminal sialic acid plays an important role in the modulation of these interactions. The expression of sialomucin genes and their glycosylation profile can be developmentally regulated, a process that acts to modulate the function of the molecule.

The major role of these two groups of molecules can be described as protection and lubrication in the case of the epithelial mucins, and as adhesion in the case of the leukocyte and endothelial cell mucin-like molecules.[4,5] Irrespective of the origin, location and biological role of a mucin, the clustering of O-glycosylation and the high content of Pro residues in the PTS regions are a common feature that has important structural consequences for the function of the molecule. The presence of Pro and the high density of glycosylated Thr and Ser residues imposes structural constraints on the polypeptide. It results in flattening of the molecule, with the PTS regions forming an extended, rod-like structure (Fig. 1).[6] Thus, oligosaccharides and their non-reducing ends are exposed to the extracellular environment and can participate in adhesive interactions. In epithelia, these structural features allow a high density of mucins to be exposed in the apical side of the cells, serving as a protective coat.

Mucin-Like Molecules in Protozoan Parasites

Mucins have been described in metazoan and protozoan parasites only recently. A number of mucins and mucin-like molecules have now been characterized and some of their genes have been identified. They vary in structure and abundance, presumably a reflection of their roles in different life-cycle stages. All of the parasite mucins and mucin-like molecules appear to be involved in some aspect of protection from the immune system and/or in host-cell invasion. Some parasites where mucin like molecules have been identified are *Toxocara canis,*[7] *Toxoplasma gondii,*[8] *Cryptosporidium parvum,*[9] *Leishmania spp.*,[10] *Trypanosoma carassii*[11] and

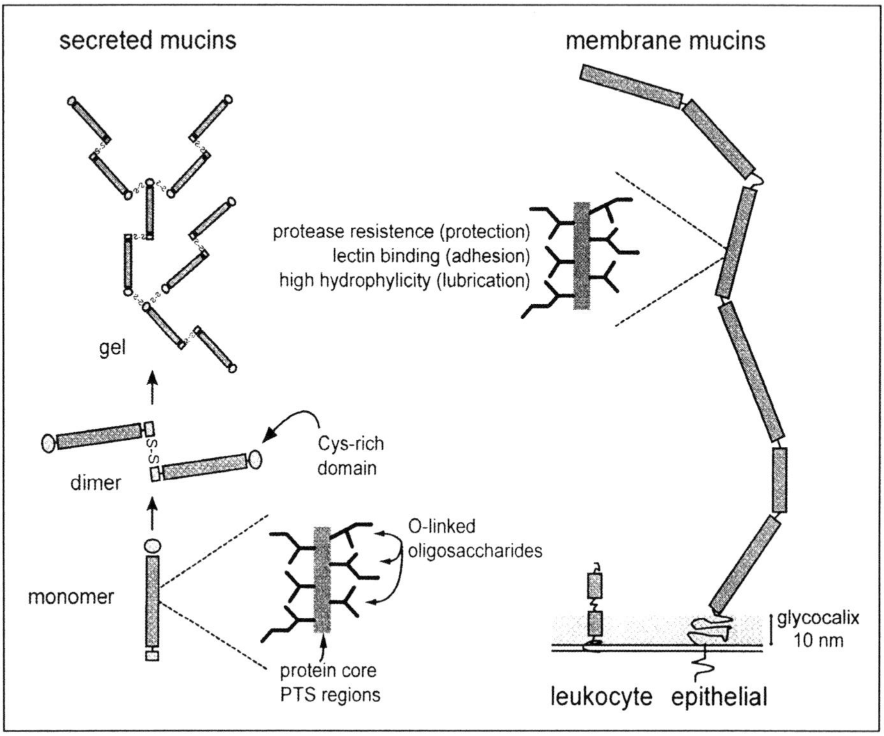

Figure 1. Scheme of the different types of mucins and mucin-like molecules from vertebrates. The main structural features of the secreted and membrane associated mucins are indicated. For the PTS (Pro, Thr and Ser rich) region, the functional implications are also shown. The membrane mucins are drawn to scale to compare the relative sizes of leukocyte and epithelial mucins.

Trypanosoma cruzi.[12,13] While in *C. parvum* the mucin described is similar to the epithelial mucins from vertebrates, having a large size of about 900 kDa and Cys-rich domains,[9] in other species mucins have unusual features. Examples include methylated trisaccharides in *T. canis*[7] and mannose bound to Ser residues through phosphate bridges in *Leishmania* promastigote filamentous proteoglycan.[10] Mucins from trypanosomes are more similar to those present in leukocytes and endothelial cells, but are anchored to the membrane by glycosylphosphatidyl inositol (GPI). In *T. carassi,* the bloodstream stage of the parasite is covered by mucins of about 36-57 kDa that are heterogeneous and highly antigenic.[11] In *T. cruzi,* mucins are present in all developmental stages and accomplish important functions for parasite survival.

Mucin-Like Molecules in *T. cruzi*

T. cruzi mucins can be broadly divided into those that are expressed in the mammalian stages of the parasite life-cycle (i.e., bloodstream trypomastigote and intracellular amastigote) and those that are expressed in the stages from the insect vector (i.e., epimastigote and metacyclic trypomastigote).

Mucins from the mammalian stages can be visualized as a 60-200 kDa smear in SDS-PAGE,[13] reflecting the fact that they are a heterogeneous population of molecules and are probably the product of different genes (see below). Special electrophoretic conditions have allowed the resolution of trypomastigote mucins as three bands with apparent molecular weights of 74, 95 and 120-200 kDa in the Y strain of the parasite.[14] Consequently, it was assumed that there may be a limited number of mucin genes expressed with the heterogeneity conferred by the different degrees of glycosylation. Each of the different bands could be identified by reactivity with a monoclonal antibody (mAb) that defined a stage-specific sialylated epitope known as Ssp3. This was shown to be formed by the action of an enzyme with *trans*-sialidase activity.[15] *Trans*-sialidase is a trypanosome-specific enzyme that transfers sialic acid from host glycoconjugates to molecules on the parasite surface. Mucins are the major acceptors of the sialic acid transferred by this enzyme.[15] The purification and analysis of trypomastigote mucins showed that they are GPI-anchored proteins which are extremely rich in Thr (20%) and Ser (10%). Half of these residues are O-glycosylated.[13] Sugars represent about 60% of the molecular mass.[13] The oligosaccharide composition has been studied and differs among strains, although the exact structure of the sugar side chains has been only partially determined. They contain 6-7 glucosidic units with GlcNAc present at the reducing end[13,16] instead of the GalNAc usually found in mucin-type O-glycosylation in most other organisms. Mucins are the most abundant membrane glycoproteins in cell-derived trypomastigotes, forming a dense coat which covers all the surface of the parasite as detected by electron microscopy and ruthenium red staining.[17]

Mucins containing the Ssp3 epitope have been implicated in mammalian cell invasion and in protection from the immune system. This followed from the observations that Ssp3 is necessary for invasion of non-phagocytic cells[18] and that passive immunization with an anti-Ssp3 mAb partially protected mice from infection.[19] Trypomastigote mucins also contain the Gal(α1-3)Gal epitope that is important in the humoral immune response against *T. cruzi* in humans.[13] Anti-Gal(α1-3) antibodies in sera from individuals in the chronic stage of the infection crosslink mucins altering the membrane organization and leading to lysis of the parasite.[17] Sialylation of the surface mucins protects the parasite from lysis caused by these antibodies.[17] A somewhat unpredicted function of the trypomastigote mucins is their role as activators of the immune system.[20] In particular, the GPI-anchor of these molecules is a very powerful proinflammatory agent.[21] The microbicidal activity of macrophages incubated with these GPI-anchors is potentiated to kill intracellular parasites.[22] Therefore, trypomastigote mucins can be both protective and detrimental to the parasite during an infection, although the relative significance of both in vivo is unknown. Mucins also seem to be present on the surface of the intracellular amastigote stage where they appear to have a similar structure to those expressed by trypomastigotes.[23] However a detailed structural analysis of these molecules has yet to be undertaken.

In the parasite stages present in the insect vector, epimastigotes and metacyclic trypomastigotes, mucins were identified as one to three broad bands that migrate in the range of 35-50 kDa in SDS-PAGE, that are periodic acid-Schiff positive.[12,24] These glycoproteins have a sugar content of about 60%, mostly in the form of O-glycan oligosaccharides. Their core proteins were found to be very rich in Thr residues (up to 30%) and to be GPI-anchored to the surface membrane.[12] The real molecular mass of these glycoconjugates was later determined by mass spectrometry to be from 13.4 to 18 kDa.[25] There were differences in the number and mass of bands detected depending on the parasite strain. As was the case with the mucins present in the mammalian trypomastigote stage, the 35-50 kDa mucins were defined and analyzed using mAbs, some of which were able to partially neutralize the infection of Vero cells by metacyclic trypomastigotes.[24] The use of several mAbs, recognizing different epitopes, demonstrated polymorphism among different strains of the parasite.[26] The structure of the oligosaccharides and the GPI-anchor of these molecules, purified from epimastigotes[27-29] and metacyclic trypomastigotes,[30] has been determined. The oligosaccharide side chains are 3-4 glucosidic units long, with some unsubstituted GlcNac present at the reducing end.[28,29] Some of the interstrain polymorphism observed using mAbs can be explained because they recognize sugar epitopes, such as Gal*f* that can be present or absent in a given strain.

The 35-50 kDa mucins are also major surface acceptors of sialic acid[12] and have been suggested to function as adhesion molecules involved in the attachment of the metacyclic trypomastigote to mammalian cells. These molecules, purified from aqueous phenol extracts of metacyclic trypomastigotes, were shown to block the invasion of mammalian cells by metacyclic trypomastigotes in a dose-dependent manner. This was not the case when the molecules were isolated from epimastigotes.[31] However understanding the roles of these mucins has not been straightforward. It has also been observed that there is a negative correlation between the abundance of 35-50 kDa mucins and infective capacity of parasite strains.[32] Moreover, treatment of the parasite with neuraminidase has been shown to favor the interaction of the parasite with non-phagocytic cells.[32] Thus, further work is required to fully elucidate the role played by 35-50 kDa mucins in adhesion and/or invasion of cells in the vertebrate host.

Some differences have been found between the 35-50 kDa mucins expressed by epimastigotes and metacyclic trypomastigotes. Epimastigote mucins have a slightly slower migration in SDS-PAGE [24] and bind much less and in a weaker manner to mammalian cells.[31] In addition they are not detected by a polyclonal serum raised against total metacyclic antigens.[24] However, the only structural difference that has been found between mucins from both stages is in the lipids of the GPI-anchor.[30] While the GPI-anchor in epimastigote mucins contains alkylacylglycerol, after differentiation to metacyclic trypomastigotes alkylacylglycerol is replaced by ceramide, a change that might modulate mucin shedding. Otherwise, the amino acid composition and the structure of the O-oligosaccharides were found to be identical in both stages. A mucin-like antigenic complex closely resembling the 35-50 kDa was reported to bind to phagocytic cells. These molecules, which migrate as 35, 45 and 50 kDa bands, were named AgC10 because they share a mAb defined epitope. They also bind to L-selectin, which is present in the membrane of macrophages.[33] Although the source and purification protocol were very similar, the AgC10 bands have been reported to be expressed in all the stages of the parasite life-cycle, unlike the 35-50 kDa mucins that are absent from the mammalian stages. Therefore at the time of writing, although there is some evidence that mucins may have a role as adhesion molecules, this remains inconclusive. Furthermore the precise nature of mucins expressed at different life-cycle stages remains to be fully determined.

The above summary of current knowledge of the structure and function of mucins in the mammalian and insect associated stages of *T. cruzi* highlights the fact that the starting material used for analysis in these studies is always complex and composed of different molecular species. Given the different experimental systems used, some of the discrepancies in the interpretation of mucin functions could be attributable to differences in the mechanisms by which *T. cruzi* infects phagocytic or non-phagocytic cells. Also, a major problem in studying the function of the *T. cruzi* mucins seems to be that, although mucins from a particular stage may share some mAb-defined epitopes, this does not guarantee that they are structurally similar and/or functionally equivalent. Furthermore, most mucin compositional and structural data that has been obtained is usually the average of many kinds of molecules. Therefore although mucins appear to be important to the parasite, it has been difficult to identify their precise function(s) using traditional immunological and biochemical approaches.

Over the last few years significant advances in the characterization of the genes that encode *T. cruzi* mucins have helped to clarify their function in the parasite. Two major families of mucin-like genes, and a third class comprising of a single member have now been described in *T. cruzi.* [34,35] Our current understanding of the nature and complexity of these mucin gene families is outlined in Table 1 and explained in detail in the subsequent sections of this review.

Table 1. ***Summary of the knowledge about the two described*** **T. cruzi** ***mucin gene families. Expression is indicated as detected by Northern blotting in the case of mRNA and Western blotting in the case of protein, for each parasite stage (i.e., E, epimastigote, M, metacyclic trypomastigote, T, trypomastigote, A, amastigote)***

Genes	Group	Product	Structural Features	mRNA				Protein				Regulation of the Expression
				E	M	T	A	E	M	T	A	
TcMUG												
500-700	I	Mucins (part of the 60-200 kDa mucins)	- O-glycosylated T8KP2 repeats - N-glycosylation site(s) - HV region at N-terminus	+	+?	+++	+	-	-	+	?	- Post-transcriptional? - Genes in tandem and scattered in genome - No AU-rich regions in UTR regions
	II	?	- No repeats (Ser/Thr rich) - Var able central region	+	+	+	+	?	?	?	?	- ? - No AU-rich regions in UTR regions - mRNA level throughout the life cycle vary for each member
	III	gp 20 kDa	- 10 putative O-glycosylation sites - Small protein core (4 kDa)	+	++	+	+	-	-	+	-	- ? - AU-rich regions in 3' UTR - Single copy gene
TcSMUG												
70-80	S	Mucins (part of the 35-50 kDa mucins)	- Small 6-7 kDa protein core - No variable regions - Thr runs with several O-glycosylation sites - One N-glycosylation site	++	+	-	-	+	+?	-	-	- Post-transcriptional by mRNA stability - AU rich regions in 3' UTR - Genes arranged in tandems
	L	?	- Small (~7 kDa protein core) - No variable regions - KNT7ST3(K/S)AP repeats - One N-glycosylation site	+++	+	+	++	?	?	?	?	- Post-transcriptional by mRNA stability - AU rich regions in 3'UTR - Genes arranged in tandems

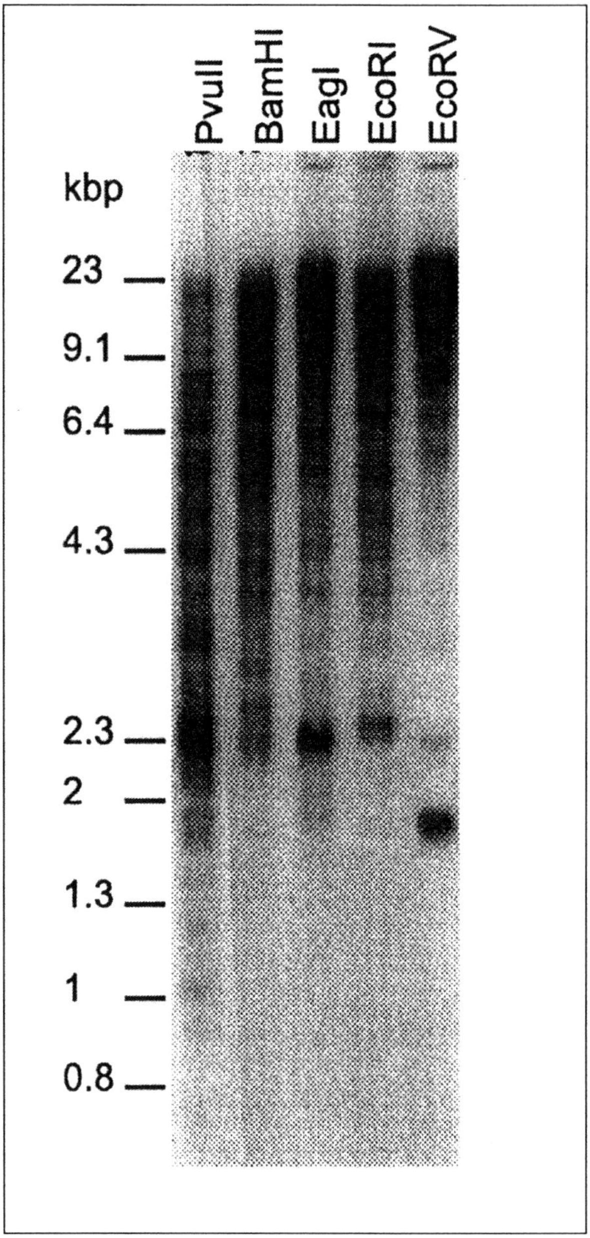

Figure 2. Complexity of the *Tcmuc* mucin-like gene family. The genomic DNA of the CL-Brener cloned stock of *T. cruzi*, digested with the indicated enzymes, was resolved on an agarose gel and transferred to membranes. Hybridization at high stringency conditions with a probe from a conserved region of the *Tcmuc* genes is shown. Size markers are on the left.

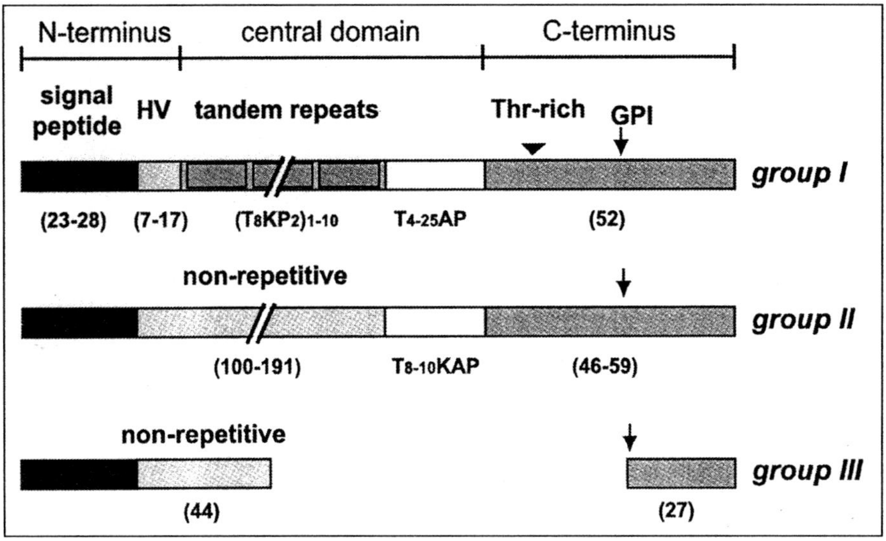

Figure 3. Structure of *Tcmuc*-deduced products. The schematic representations of the TcMUC-deduced products are shown. The different regions are defined and named. The numbers in brackets below each region indicate the length range in terms of amino acids. The inverted triangle indicates a consensus N-glycosylation site conserved in the group I products. The arrow indicates the position of the probable GPI-anchor addition site.

Tcmuc: A Complex and Highly Diverse Mucin-Like Gene Family

The first mucin-like gene from *T. cruzi* was isolated from a genomic library by immunological screening with a human infection serum.[36] Using probes derived from this sequence, Southern analysis revealed that the gene was part of a family with unsuspected complexity (Fig. 2). Many genes belonging to this family (named *Tcmuc* after the mammalian mucin genes), have now been cloned from different *T. cruzi* strains,[25,34,37,38] in particular CL-Brener, the genome project reference strain. The sequencing of over 60 clones, combined with hybridization against genomic Southern blots and a high-density arrayed cosmid library, provided an estimate of at least 480 genes per haploid genome.[39] In keeping with this, a recent survey of over 10,000 random genomic DNA sequences estimated a gene copy number of 710 for CL-Brener.[40] *Tcmuc* genes are highly diverse and can be classified in three groups (I-III) whose deduced products and main features are depicted in Figure 3.

The *Tcmuc* group I is composed of genes encoding small proteins with an average of 150 amino acids. They have an endoplasmic reticulum import signal at the N-terminus. This is followed by a central domain composed of a variable number of head-to-tail tandem repeat elements with the consensus sequence T_8KP_2. The C-terminal region of the protein contains a GPI-anchor addition signal. The functionality of the signal peptide and the GPI anchor signal were demonstrated following transfection of a tagged version of a group I gene into *T. cruzi* epimastigotes.[41] The expressed protein was highly O-glycosylated with some N-glycosylation and it was GPI-anchored to the surface membrane. Results from Almeida *et al* suggest that in trypomastigote mucins an Asp residue located near the C-terminus and conserved in each of the three groups (Fig. 3) acts as the GPI addition site.[21] The SDS-PAGE mobility of the transfected product, the fact that the T_8KP_2 repeats were recognized by 90% of sera from infected mice and the mRNA profile determined by Northern blotting, indicate that this group I gene encodes a core protein similar to the mucins from cell-derived trypomastigotes.[34,37,41] The most interesting structural feature of the group I proteins is the presence of a hypervariable region (HV) (Fig. 3). The relevance of this will be addressed below.

The deduced products of the group II mucin genes have signal peptide and C-terminal regions similar to the group I proteins. They share an average of 70% sequence identity. However, the central domain of the group II proteins contains unique regions that are distinct, not only from the T_8KP_2 repeats but also from each other (Fig. 3). This increases significantly the diversity within the family. Between this central region and the C-terminus, this group of mucins also contains one or two degenerate repeat elements, with a sequence similar to T_8KP_2. Unfortunately we do not yet have many clues about the nature of the product encoded by these genes. Transfection-mediated expression in epimastigotes did not yield a highly glycosylated product, although we did find evidence of post-translational modifications, which are as yet undefined.[41] The stage-specificity of group II mRNA expression was studied but did not give a definitive answer. The mRNA levels varied for individual genes at different developmental stages of the *T. cruzi* life cycle.[37,39] When the variable central domains of some group II genes were aligned and compared, they were found to have similar numbers of synonymous and non-synonymous mutations. It has been proposed that these sequences evolve neutrally, without selective pressure and that this region is removed from the mature molecule and not expressed on the cell surface.[37] As yet there is no conclusive evidence to support or negate this hypothesis.

The group III mucin is defined by a single copy gene that differs from groups I and II. It has a short non-repeated central region but no Thr runs (Fig. 3). It does however have a signal peptide and GPI-anchor signal similar to the other two groups, allowing its classification as member of the *Tcmuc* family. The rest of the deduced protein, representing the mature peptide, is only 44 amino acids long with a molecular weight of 3800 and 10 predicted O-glycosylation sites, but no consensus N-glycosylation site. The existence of this product as a GPI-anchored protein in the membrane of the trypomastigote stage was demonstrated using antibodies raised against the recombinant protein expressed in bacteria (our unpublished results). The natural product migrates at 20 kDa in SDS-PAGE and therefore may undergo post-translational modification. It is highly antigenic during the infection in humans, mice and rabbits. This is the only TcMUC protein that has so far been found to have a completely non-repeated central region and that is expressed in vivo.

Hypervariable Regions in *Tcmuc* Genes

As mentioned above, the sequence of the group I mucins between the signal peptide and the T_8KP_2 repeats is hypervariable (HV). This HV region is polymorphic among the members of the group I mucins in terms of both sequence and length, which can range from 8 to 17 amino acids. Of 32 cDNA and genomic clones that we analyzed from CL-Brener, 22 variants were found. The mutations that produce this hypervariability are region-specific in the gene. Alignment of 400 bp of 5' non-coding region and the sequences encoding the signal peptide indicates at least 90% identity between genes. However the sequences of the HV region and the tandem repeats are so diverse that they are difficult to align. The number of repeats can differ between genes and there is also a Thr-rich stretch (TnAP) in which the number of Thr residues can vary from 6 to 25. One possibility is that this region can undergo recombination between members of the gene family and that the differences in the numbers of repeats and Thr codons could arise from unequal crossover. Notwithstanding this, almost all of the mutations found in the repeats of the gene are synonymous, as is the case for sequences encoding the N- and C-termini. The 300 bp of sequence that constitute the 3' untranslated region (UTR) of the mature mRNA are highly conserved within group I and share about 90% identity. In contrast to the rest of the coding or non-coding regions of the group I genes, in the HV region every mutation identified was non-synonymous and, in many cases, not even conservative.

```
     Cys6                                                       Cys6
 1   TGCGTGACAGC--AAG-TGAT------AAG----------CAGGAA---act       CVTASD--K-----QET
 2   TGCGTGACAGC--AAG-TGTT------AAG----------CAGGAC---act       CVTASV--K-----QDT
 3   TGCGTGACAGC--AAG-TGAT---CAGAAG----------CAGGAC---act       CVTASD-QK-----QDT
 4   TGCGTGACAGC--AAG-TGTT---CAGAAG----------CAGGAA---act       CVTASV-QK-----QET
 5   TGCGTGACAGC--AAG-TGGT---CAGAAG----------CAGGAA---act       CVTASG-QK-----QET
 6   TGCGTGACAGC--AAG-TGGT---CAGAAG----------CAGGAC---act       CVTASG-QK-----QDT
 7   TGCGTGACAGC--AAG-TGGT---CAGAAG---GCTGAG-CAGGAC---act       CVTASG-QK---AEQDT
 8   TGCGTGACAGC--AAG-TGGC---CAGAAG---GCTGAG-AAGGAA---act       CVTASG-QK---AEKET
 9   TGCGTGACAGC--AAG-TGGTGATCAGAAGCAGGCTGAG-CAGGAA---act       CVTASGDQKQ--AEQET
10   TGCGTGACAGC--AAG-TGGT---CAGGAG---GCTAAG-CAGGAA---act       CVTASG-QE---AKQET

11   TGC-----AGCA-AAGAGGCTGTTGAGAAGAAGGCTGAG-CAGGAA---act       CSKEAVEKK---AEQET
12   TGC-----AGCG-AAGAGGGT------AAG----------CAGGAA---act       CSEE--------GKQET
13   TGC-----AGCG-AAGAGGGT------AAG----------CAGGAC---AATacc    CSEE--------GKQDNT
14   TGC-----AGCG-AAGAGGTT------AAG----------CAGGAC---AATacc    CSEE--------VKQDNT

15   TGCGCGGCAG---AAGGTGGTGGTCAGAAG----------CAGGAA---AATacc    CAAEGGGQK-----QENT
16   TGCGTGACAGCG-AA--TGGT--AAAGAAGGA-TCCGAG----------acc       CVTANG--KEG---SET
17   TGCGTGACAGCG-AA--TGCT--AA-----------------TTCTTCCAGTAACaca CVTANAN------SSSNT
18   TGCGTGACAGCG-AA--TGGC--AAAGAAGGA-----AA-TCCAGTTCCaca       CVTANG--KEG-ESSST
19   TGCGTGGCAG---------ATGCTCAGCCTGTGCCTGAAGGG-TCTTCCAATacg    CV---ADAQPVPEGSSNT
20   TGCGTGGCAG---------ACGCTCAGTCTGTGCCTGAAGGG-TCTTCCAATacg    CV---ADAQSVPEGSSNT
21   TGCGTTTCAGAGGAAGCTGCCGCTCAGCCTGTGCCTGAAGGG-TCTTCCAATacg    CVSEEAAAQPVPEGSSNT
22   TGCGCGTCAG-GCACTGAGCCTGAGCCTGAAGCTGCCGCTTCTTCTTCCAATacg    CASGTEPEPEAAASSSNT
```

Figure 4. Hypervariable regions of *Tcmuc* group I genes. Alignment of DNA (left) and the deduced amino acid (right) sequence of the HV region from the group I *Tcmuc* genes. Sequences shown are all the variants found from CL-Brener by sequencing 32 cDNA or genomic clones. A conserved Cys codon and the first Thr codon from the first repeat (in lowercase) were taken as the boundaries of the sequence aligned. Nucleotides different from the consensus are shaded. Sequences are grouped based on the amino acid sequence similarities. Alignment was performed manually; dashes are spaces introduced for best alignment.

Alignment of all known HV regions from CL-Brener group I genes did lead to the detection of some similarity, suggesting that the variation has resulted from gene duplication and mutation accumulation (Fig. 4). In summary, numerous non-synonymous mutations are clustered in the HV region of group I genes, whereas the rest of the sequence has accumulated considerably less mutations, with a major proportion of these being synonymous.

For group II genes, the whole non-repeated central region can be considered as HV (Fig. 3). However, when several of the deduced peptide sequences were aligned, some similarities were observed in this region (Fig. 5). Most of these were located towards the C-terminal region of the variable domain. Almost no similarities were identified in the region close to the signal peptide, analogous to the HV region of group I. A clear example of this is shown in Figure 6A. An additional observation was the finding that two cDNA clones, derived from mature mRNA species that had undergone splice leader and polyA tail addition, had coding regions interrupted by stop codons.[39] A group II genomic clone with a truncated reading frame has also been isolated (our unpublished results) and another is reported in GeneBank (accession number AJ239062), both from CL-Brener. A cDNA clone with an internal stop codon has also been isolated from the Y strain of *T. cruzi*.[37] In every case these internal stop codons are localized within the HV region or the variable domain (Fig. 6B).

This evidence suggests that mechanisms exist allowing the HV region to act as a mutational hot spot. The presence of degenerate repeats at the end of the variable region of the group II proteins can be taken as evidence that group I proteins containing the T_8KP_2 repeats were the ancestors of group II, and that these have since diversified by the accumulation of mutations. The existence of a polarity in mutation accumulation, from 5' to the 3' end, and the isolation of transcribed pseudogenes, points to an intrinsic mechanism that allows the preferential

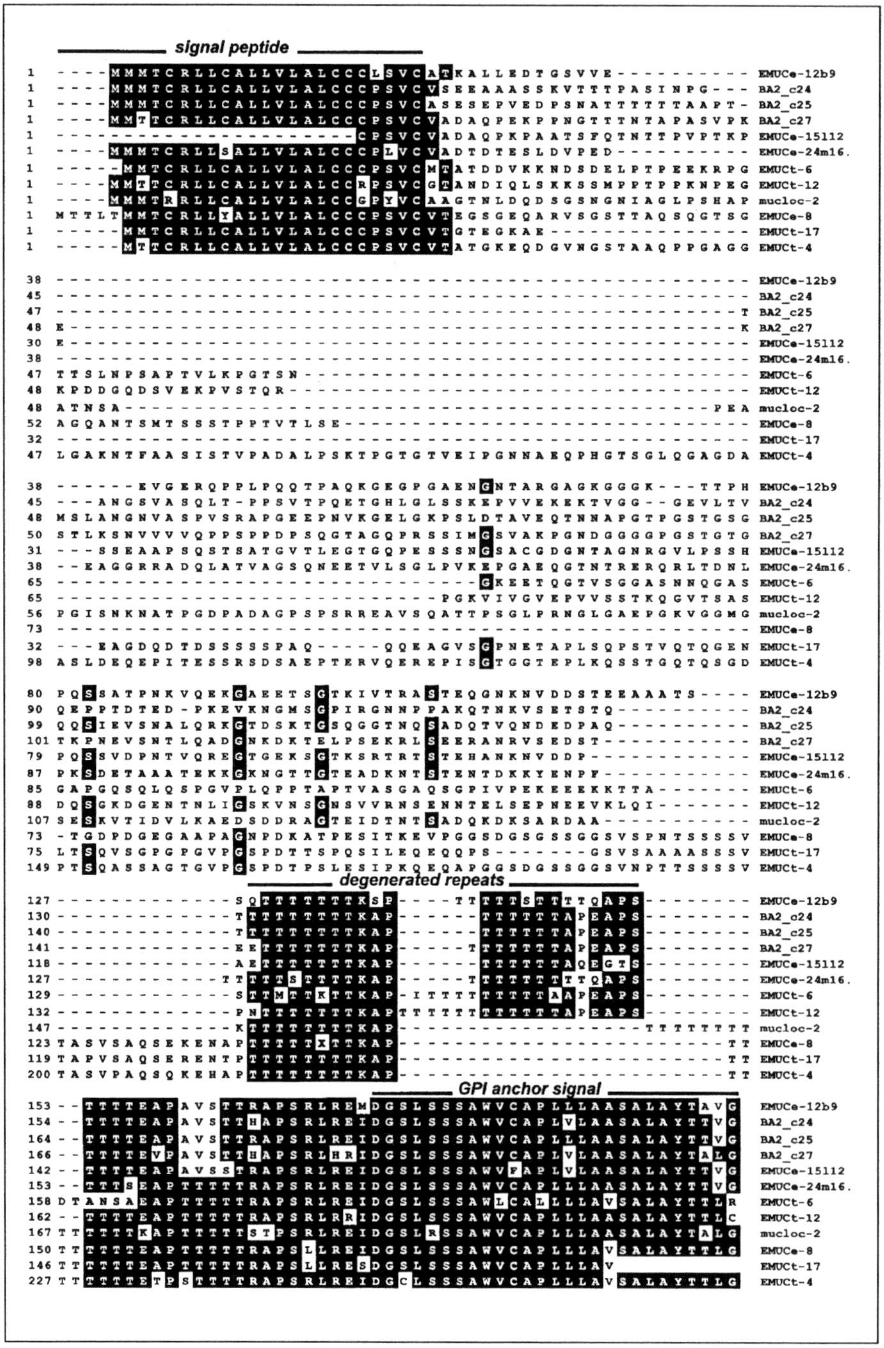

Figure 5. Alignment of proteins deduced from group II *Tcmuc* genes. All the sequences come from complete genes from CL-Brener clone. Alignment was performed by the CLUSTAL method using MegAlign Software (DNASTAR Inc.) Black shading indicates identities among six or more sequences. Dashes are spaces introduced for best alignment.

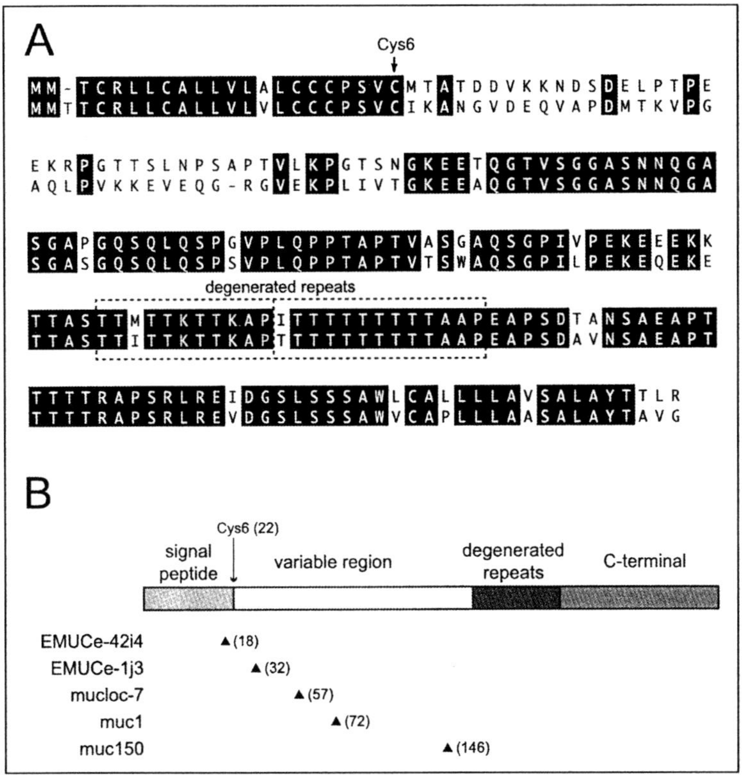

Figure 6. Evidence that mutations are preferably localized within the variable and hypervariable region of *Tcmuc* genes. Panel A, Two deduced proteins from *Tcmuc* group II were aligned. Black shading indicates identities. Dashes are spaces introduced for best alignment. The Cys 6 is indicated as the 5' boundary of the variable region. Panel B, Schematic representation of the pseudogenes cloned from the *Tcmuc* family. All the indicated clones belong to group II. Triangles indicate the position of stop codons (codon number in brackets). The first three schemes correspond to CL-Brener sequences from ref. 39. Unpublished sequences obtained from GeneBank have the following accession numbers *muc150* AF027876 (ref. 37) (Y strain), *muc1* AJ239062 (CL-Brener strain).

accumulation of mutations in a site-specific manner. Natural selection may also participate in this process as will be discussed below.

The Second Mucin-Like Gene Family from *T. cruzi*: *Tcsmug*

A new group of mucin-like genes, distinct from the *Tcmuc* family, has more recently been detected in *T. cruzi*.[35] These have been designated the *Tcsmug* family; for Small Mucin-like Genes given their size, amino acid composition and the structure of the deduced peptides (Fig. 7). This family is composed of two groups of genes, named L and S for Long and Short, respectively, based on the length of their mRNAs. The deduced proteins from both groups are highly similar in the N- and C-terminal regions, the signal peptide domain and the GPI-anchor signal region respectively. Both groups have an N-glycosylation consensus sequence adjacent to the GPI addition site (Fig. 7). The central domains of both groups contain Thr-rich sequences. In the case of the L group, Thr residues are located within tandem repeats having the consensus sequence KNT$_7$ST$_3$(K/S)AP. For the S group, the central region has the consensus sequence DQT$_{17\text{-}20}$NAPAKDT$_{5\text{-}7}$NAPAK. Consistently in the S group, the first stretch of Thr

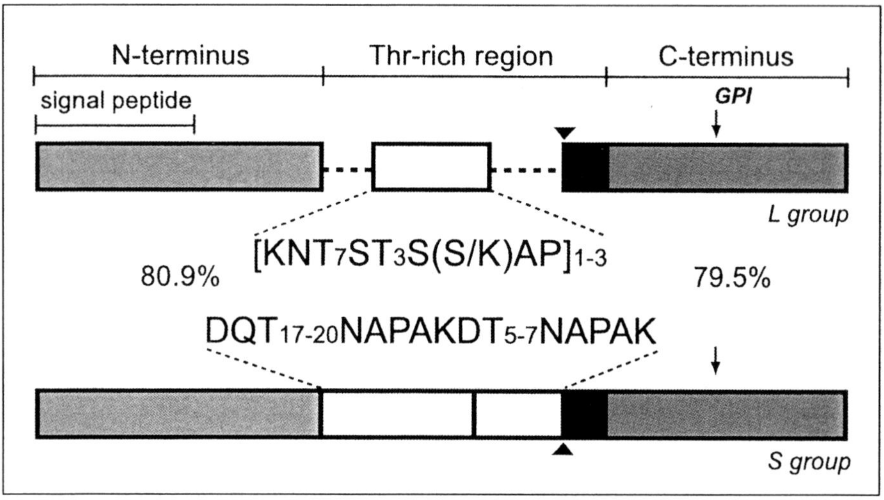

Figure 7. Structure of the TcSMUG-deduced products. The regions of the deduced products from the L and S group are schematized. The inverted triangles indicate a conserved consensus N- glycosylation site. The arrow indicates the predicted GPI-anchor site. The consensus sequence of the central regions is indicated in each case. Percentage of identity for each conserved region[s] is indicated. The graphic is not to scale.

residues is longer than the second, which on a few occasions has been observed to be duplicated or absent. There are no variant regions in these genes, although there is small subset within the S group varying at four positions. Therefore the differences between members of this family are conferred by variation in the number of repeats in the case of the L group, and by heterogeneity in the number of repeated Thr residues in the case of the S group. These differences may have originated by recombination.

The *Tcsmug* genes are organized in tandem repeats in the genome. The arrays characterized so far contain only S or L genes, but not both.[35] The family has around 70 genes per haploid genome for the CL-Brener clone. Even though this can vary between strains, there is clearly far less complexity than within the *Tcmuc* family, as can be judged from Southern blots (Fig. 8A). Members of the L and S groups differ in their mRNA expression pattern. The L group is expressed throughout the parasite life cycle, but is more abundant in the replicative stages (epimastigote and amastigote). The S group mRNA level is maximal in epimastigotes, less in metacyclic trypomastigotes and very low in amastigotes and cell-derived trypomastigotes (Fig. 8B). The extragenic regions of these genes have been studied [35] and the *trans*-splicing and polyadenylation signals determined. The 5' UTR of all the transcripts share 90% identity. The 3' UTR of S and L transcripts have a modular organization with some boxes conserved in both, while other boxes are exclusive to one or the other group (Fig. 8C). The expression is post-transcriptionally regulated at the level of mRNA stability and some of the boxes in the 3'-UTR have been identified as *cis*-acting elements involved in this mechanism (discussed further below). These observations might explain the differential pattern of mRNA abundance throughout the lifecycle of the two groups.

The products of the S group were identified following sequencing of N-terminal peptides obtained from the 35-50 kDa mucins of epimastigotes (Dr. I. Almeida, personal communication). The sequences corresponded with those of the S group proteins, and also indicated that the mature N-terminus is different from that predicted.[35] So, further processing of the molecule ocurrs after endoplasmic reticulum import. These results, together with the similarities in molecular mass and expression patterns, indicate that at least some of the mucins expressed in

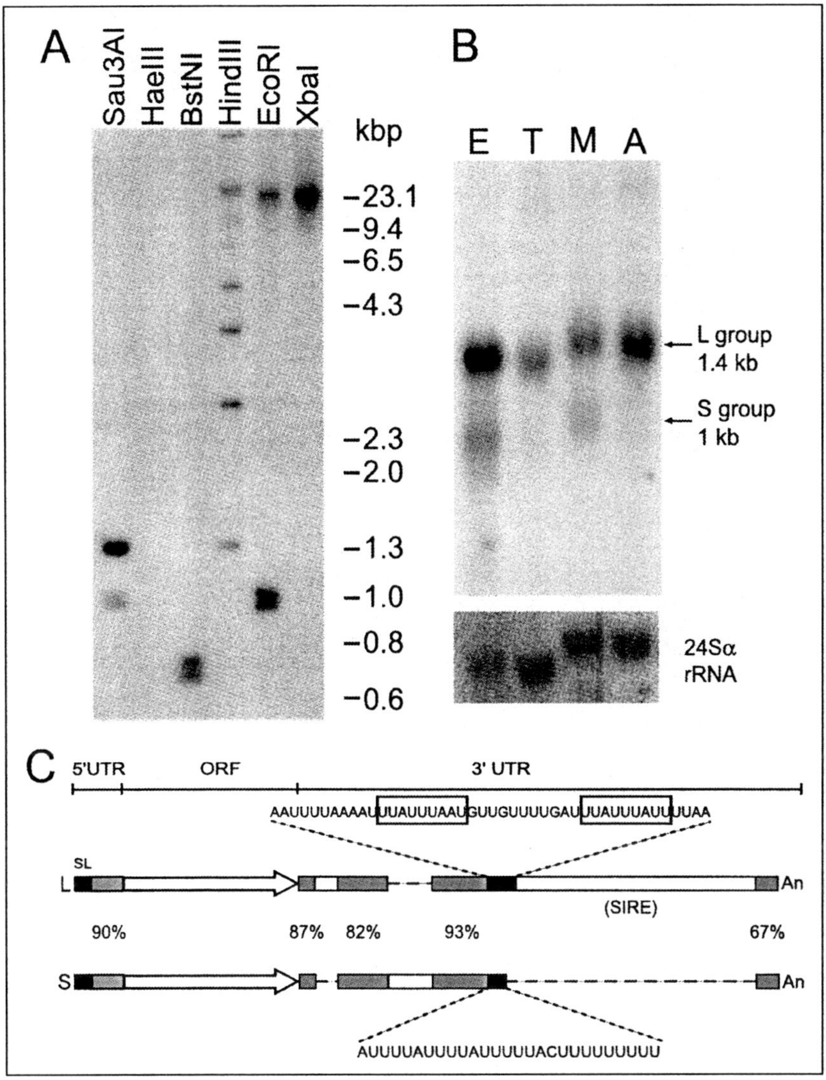

Figure 8. Genomic organization and expression of the *Tcsmug* gene family. Panel A, Southern blot analysis of genomic DNA from *T. cruzi* CL-Brener digested with the indicated enzymes and probed at high stringency with a fragment from a *Tcsmug* conserved region that hybridized to both the L and S genes. Note that the HindIII digestion was partial revealing the tandem organization of the genes. Panel B, Northern blot of total RNA from *T. cruzi* epimastigotes (E), cell-derived trypomastigote (T), metacyclic trypomastigote (M) and amastigote (A). The position and size of the L and S group transcripts are indicated. Panel C, Scheme of the L and S mature transcripts as deduced from cDNA sequencing. The conserved modules from the 3' UTR are shaded.

epimastigotes are the product of the *Tcsmug*-S genes. When the amino acid composition determined for epimastigote purified mucins[12,30] and that obtained from the deduced products of the S genes are compared, there is a discrepancy. This suggests that the product of (an)other gene(s), including possibly members of the L group, may account for the two or three bands observed in SDS-PAGE of purified epimastigote mucins.

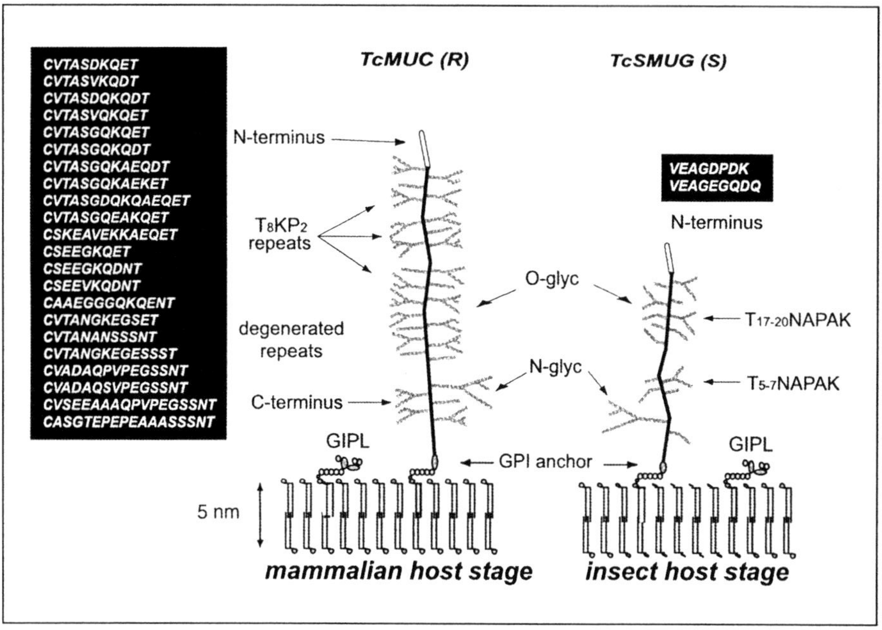

Figure 9. Structural model for epimastigote and trypomastigote mucins. The schemes show the main features and the structure for the *Tcmuc* group I and *Tcsmug* group S products, based on the available data as described in the text. Sequences in the black boxes are from the N-termini present in members of TcMUC (R) and TcSMUG (S) families.

Structure and Function of TcMUC and TcSMUG Mucins

Two groups of genes have now been experimentally shown to encode mucins in *T. cruzi*: the group I of the *Tcmuc* family, and the group S of the *Tcsmug* family (Table 1). The structure of these two glycoproteins can be schematized, as shown in Figure 9, on the basis of the experimentally determined features mentioned above, the predicted lack of secondary structure and the effect of high density O-glycosylation. The HV region is present at the N-terminus of mature molecules encoded by group I of *Tcmuc* family (ref 41 and our unpublished results). As mucins form a dense coat in trypomastigotes,[17] the HV region should be exposed on top of the stalk formed by the central domain, protruding into the extracellular medium and accessible to the immune system. The observed variability in the HV region could be a result of positive Darwinian selection as described for several antigen families.[42] The display of variant epitopes is a well-known immunoevasion mechanism that has been described for other pathogen antigens like the E2 protein of the Hepatitis C virus,[43] the circumsporozoite protein of *Plasmodium* spp.,[44] or members of the SA85 antigen family of *T. cruzi*.[45] Our working hypothesis is based on the assumption that strong antibody responses against the HV region may be harmful for the parasite. Antibodies against these regions have been detected in sera from infected mice, rabbit and humans.[41] This confirms both the presence of the HV region in the mature protein, and the fact that it is detectable by the immune system. The S group of the *Tcsmug* family encodes mucins present in the insect stages of the parasite. Insects exert only non-specific immune responses and, accordingly, epitope diversity is not required and does not occur in TcSMUG mucin core proteins that are expressed in this parasite stage. The 35-50 kDa epimastigote mucins however are relatively resistant to some protease treatments, in contrast with trypomastigote-specific mucins that are more sensitive. The gut of the insect is very rich in proteases, and so mucin subtypes might have been selected to confer protection in this hostile environment. Therefore, *T. cruzi* is able to switch expression of the molecules that form part of its surface coat in a stage-specific manner. The mechanisms involved in regulating gene expression in trypanosomatids are still poorly understood. The developmental control of the complex mucin gene family therefore provides an excellent model for studying this process.

Regulation of *T. cruzi* Mucin Gene Expression

mRNA Quality Control in Eukaryotes

Regulation of transcription initiation is the main step used by most eukaryotic cells to control gene expression. In addition, regulation of nucleo-cytoplasmic transport, mRNA stability and translational masking facilitate a coordinated regulation of the RNA maturation process. These mechanisms operate jointly to control the fate of a mRNA, and form part of the quality control machinery crucial to the regulation mRNA function.[46] The major steps of pre-mRNA processing (capping, splicing, 3'end-formation and polyadenylation) have been shown to be coupled to transcription.[47,48] After nuclear processing, and transport to the cytoplasm, the level of mature mRNA is subjected to tightly controlled turnover mechanisms that act to ensure the proper expression of the final gene product. Regulation of mRNA stability in different cell types involves two components: *cis*-elements made up of defined sequences, most frequently located in the 5'- and 3'-UTR of the mature mRNA, although some could also be present within the coding region;[49] and *trans*-acting factors which are proteins that bind to these *cis*-elements and modulate RNA metabolism. The combination of one or more *cis*-elements in a given mRNA and the type and number of *trans*-acting factors, determine whether a mRNA will be stabilized for translation or rapidly degraded by ribonucleases. Thus, protein interaction with specific RNA sequence elements represents a coordinated mechanism that modifies the fate of a mRNA.

One well-characterized *cis*-element that regulates mRNA stability is the group of AU-rich elements (ARE) found in the 3'-UTR of short lived mRNAs,[50,51] such us those of proto-oncogenes and cytokines (e.g., IL-1, 2, 3 and 10).[52-54] Each ARE represents a combination of functionally and structurally distinct sequence motives, such as the AUUUA pentamer, the UUAUUUA(U/A)(U/A) nonamer, stretches of uridines and/or U-rich domains, that can range in size from 50 to 150 bp. ARE-directed mRNA decay has been linked to cell transformation, cell growth and differentiation, cell adhesion and to the immune response.[55] An effect of the ARE on translation efficiency was also described, both positively[56] and negatively.[57] A number of RNA-binding proteins have been identified in higher eukaryotes that interact with ARE sequences. Many of these factors contain highly conserved RNA-binding domains that place them within the RRM (RNA recognition motif) superfamily.[58] *Trans*-acting factors that recognize ARE can increase mRNA stability or induce mRNA degradation.[59,60]

Regulation of Gene Expression in Trypanosomes

At variance with higher eukaryotic cells, control of gene expression in trypanosomatids is mainly post-transcriptional, predominantly at the level of mRNA maturation (reviewed in refs. 61,62). There are several examples in which post-transcriptional mechanisms facilitate expression in a developmentally regulated manner.[35,63] The importance of post-transcriptional events in the control of gene expression in trypanosomes is a consequence of two findings. First, α-amanitin-sensitive RNA polymerase II from trypanosomes transcribes large polycistronic units containing many different and functionally unrelated coding sequences.[64] This precludes regulation at the level of transcription initiation. Polycistronic precursors are then processed into mature mRNAs by coordinated *trans*-splicing and poly(A) addition on the 5′- and 3′-ends, respectively.[65] Second, conventional RNA polymerase II promoters might not be present in trypanosomes, and it has been suggested that genomic accessibility and double-strand melting are the processes involved in RNA polymerase binding.[66] Transcription initiation by RNA polymerase II may require as yet uncharacterized initiator elements-protein interactions rather than the usual binding proteins specific for the TATA box.[67]

Possible post-transcriptional processes that can be the subject of regulation in trypanosomes include *trans*-splicing, nucleo-cytoplasmic transport, mRNA stability and translational control. Most recent work points to regulation of mRNA stability as one of the major mechanisms. However, few *cis*-regulatory elements have been identified. Of these, most are located in the 3'UTR and act by altering the half-life of mature mRNAs.[35, 63] One example is the 16-mer loop localized in the 3'UTR of procyclic mRNAs of *T. brucei* that confers stage-specific mRNA stability and improves translation efficiency.[68] Different 3'UTRs and intergenic regions have been shown to influence expression by changing the steady-state level and/or the translation efficiency. The presence of U-rich regions and the length of the 3'UTR have been demonstrated to regulate mRNA polyadenylation and translation efficiency of a reporter gene in trypanosomes.[69] However, almost no information is available on the *trans*-acting factors that might be involved in the process of regulation of mRNA expression.

Post-Transcriptional Control of Tcsmug *Gene Expression in* T. cruzi

As mentioned above, the *Tcsmug* mucin gene family is developmentally regulated. The mRNA steady-state level is higher in the epimastigote than in the metacyclic trypomastigote stage of the life cycle in the case of both the L and S transcripts (see Fig. 8B). However, the rate at which the genes are transcribed is the same in both parasite stages.[35] Thus, *Tcsmug* gene expression appears to be regulated primarily at the post-transcriptional level. Using the RNA polymerase inhibitor actinomycin D (ActD) we have shown that the half-life of these transcripts is more than six hours in the epimastigote stage. Conversely, they are short-lived in the trypomastigote stage of the parasite, having a half-life of less than 30 minutes.[35] These results demonstrated a clear correlation between mRNA steady-state levels and half-lives during parasite development, reinforcing the importance of post-transcriptional mechanisms in the control of mRNA abundance. In vivo treatment of parasites with the protein synthesis inhibitor cycloheximide (CHX) did not seem to exert any significant effect on mRNA levels. However, the treatment of epimastigotes with both CHX and ActD caused a three-fold reduction in the half-lives of the *Tcsmug* transcripts.[35] One explanation for this reduction might be the downregulation of a labile positive factor due to protein synthesis inhibition. However, the presence of CHX alone did not modify the half-life of the trypomastigote *Tcsmug* transcripts. Therefore, the instability in this parasite stage seems not to be dependent on translation, as is the case of several RNAs bearing AU-rich elements in higher eukaryotic cells.[70,71] There are two simple explanations for this behavior. One is that these mRNAs are intrinsically short-lived and that a mechanism to protect them from degradation, mediated by some labile factor(s), could be acting in epimastigotes and not in trypomastigotes. Alternatively, the mRNA could be protected while being translated and then become short-lived in the absence of translation. Whichever of these holds true, it is clear that the mRNA of *Tcsmug* S genes are stabilized in epimastigotes, the stage of the parasite where they are highly expressed and not in trypomastigotes, where mucins from the other gene family (*Tcmuc*) are preferentially expressed.

3'UTR cis-*Elements Controlling mRNA Stability and Translation Efficiency of* Tcsmug *Transcripts*

A number of *cis*-elements involved in the control of *Tcsmug* mRNA stability and translation throughout the parasite life cycle were recently identified using deletion mutants fused to a *cat* reporter gene (Fig. 10).[35,72] A 44 nt AU-rich element (ARE) present in the 3'UTR was

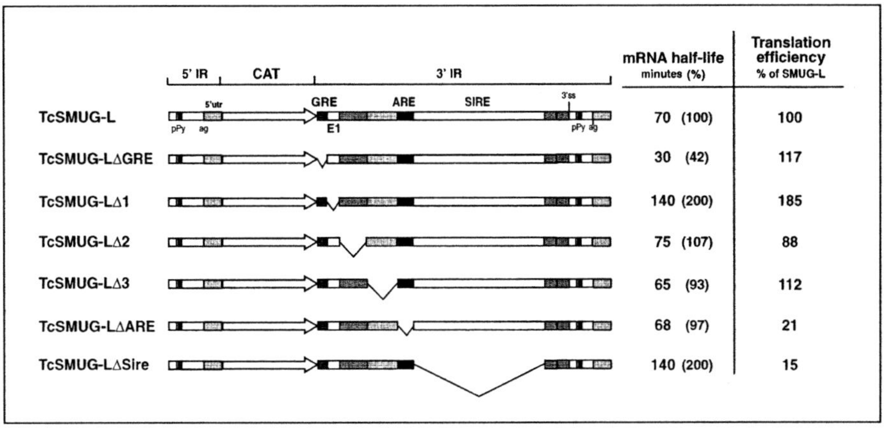

Figure 10. Identification of positive and negative *Tcsmug*-L 3'UTR *cis*-elements involved in mRNA stability and translation modulation. Schematic representations of complete *Tcsmug*-L and 3'UTR deletion mutants are shown. All constructs were obtained by PCR as described previously.[72] The 5' and 3' intergenic regions (IR) contain the original *trans*-splicing site (ag) and polypyrimidine tract (pPy) for efficient mRNA processing. Epimastigote forms of the parasite were transfected with the indicated DNA constructs cloned in the pTEX vector.[76] After selection of recombinants, parasite populations were treated with ActD and Northern blots were performed to estimate the half-life of the transcripts derived from each construct. A CAT activity assay was conducted and compared to the mRNA steady-state levels of each transcript. Values of translation efficiency (expressed as % of *Tcsmug*-L activity level) are indicated on the right side of the panel.

shown to destabilize *Tcsmug* transcripts in the metacyclic trypomastigote stage but not in epimastigotes.[35] A second *cis*-element that stabilizes *Tcsmug* mRNAs in the epimastigote stage was identified as a 27 bp G-rich element (GRE). This has been designated GRE and is composed of two contiguous CGGGG pentamers. A third element, located 28 and 62 nucleotides downstream of the stop codon (named E1) was found to have a negative effect on mRNA stability in the epimastigote stage. Its removal from the 3'UTR of *cat* mRNA increased the half-life of this reporter transcript up to two-fold (Fig. 10).[72] Some of these elements were also shown to affect translation efficiency of *cat* mRNA (Fig. 10). This phenomenon has previously been reported with higher eukaryotic cells.[35]

In brief, there are positive and negative *cis*-elements within the 3'UTR of *Tcsmug* mucin transcripts, which regulate mRNA stability and translation efficiency. The effect on mRNA stabilization is, in some cases, stage specific. Thus, some *cis*-elements, like GRE and E1 affect mRNA stability in the epimastigote stage, while the ARE sequence was observed to function in the metacyclic trypomastigote stage (Fig. 11). Furthermore, one element stabilizes mRNA (GRE) while others are involved in mRNA destabilization (ARE and E1) (Fig. 11). These results, together with the finding that the treatment of epimastigotes with CHX caused a reduction in the half-lives of *Tcsmug* transcripts,[35] indicates that functional *trans*-acting factors may be involved in interacting with these elements.

Trans-*Acting Factors Interacting with* Tcsmug *mRNAs: ARE and GRE RNA-Binding Proteins*

Proteins binding to the *cis*-elements involved in the regulation of the *Tcsmug* mRNA stability were identified by electrophoresis mobility shift assay (EMSA), using ARE and GRE RNA sequences and total *T. cruzi* protein extracts (Fig. 12).[72] Different ribonucleoprotein complexes were detected with both RNAs. The AU-rich element RNA-binding proteins

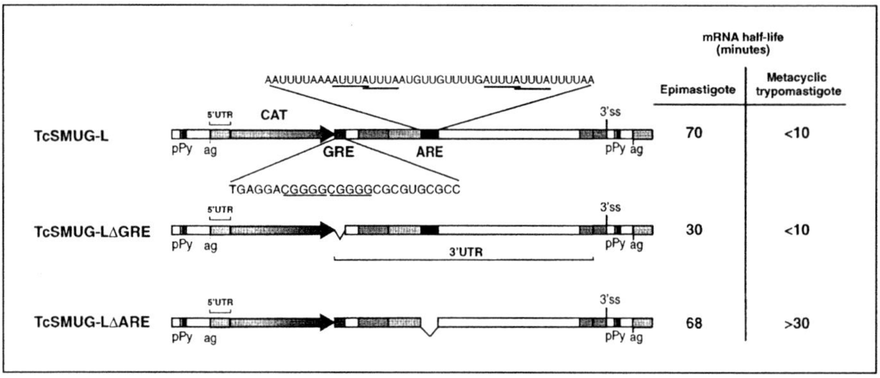

Figure 11. Two *cis*-elements, AU- and G-rich, conferred a functionally different, developmentally regulated expression pattern throughout parasite life cycle. Schematic representation of *Tcsmug*-L (complete construct), *Tcsmug*-L-GRE and *Tcsmug*-L-AU deletion mutants used to transfect the epimastigote stage of the parasite. The sequence that was deleted in clone *Tcsmug*-L-GRE and *Tcsmug*-L-AU is indicated in the *Tcsmug*-L scheme. After selection of recombinants, parasite populations were treated with ActD and Northern blots were performed to estimate the half-life of the transcripts derived from each construct in two different parasite stages, epimastigote and metacyclic trypomastigote. The half-life (expressed in minutes) of each transcript is indicated on the right side.

(ARE-BPs) were found to differ during the parasite life cycle. The epimastigote ARE binding protein (E-ARE-BP) migrated much more slowly in a native gel than the complexes detected in the other three parasite stages. By UV cross-linking analysis we demonstrated that E-ARE-BP has a larger apparent molecular mass (about 100 kDa) than those detected in the other parasite stages (about 45-50 kDa). Conversely, GRE RNA migrated as part of the same three ribonucleoprotein complexes (named G-complexes) in all parasite stages (Fig. 12B). The G-complex-1 gave rise to a single band having an apparent molecular mass of 80 kDa, while the G-complexes-2 and -3 were found to be composed of several proteins with apparent molecular masses ranging from 35, 39 to 66 kDa. The minimal sequences required for the binding of ARE- and GRE RNA-binding proteins were determined using a set of oligoribonucleotides. The sequences were the entire 44nt AU-rich sequence for E-ARE-BP and two contiguous CGGGG pentamers for GRE-binding proteins.[72]

Subcellular fractionation demonstrated that E-ARE-BP is mainly cytoplasmic and might be partially associated with polysomes, while T-ARE-BPs (trypomastigote ARE binding proteins) are localized in both the nucleus and the cytoplasm, and might function as nuclear-cytoplasmic shuttling RNA-binding proteins. Similar experiments performed with G-complex-1 and G-complex-2 indicated that the former is localized in the cytoplasm while the latter is equally distributed between the nucleus and cytoplasm. Consequently, it is possible that those GRE RNA-binding proteins might protect the mRNA during transport into the cytoplasm. Several proteins in higher eukaryotes and also in trypanosomes have been shown to exhibit a shuttling behavior between nucleus and cytoplasm.[73,74]

We have recently cloned members of a large family of genes encoding U-rich RNA-binding proteins that are developmentally regulated (our unpublished results). They share several structural features with RNA-binding proteins from higher eukaryotic cells and recognize defined *cis*-elements on *Tcsmug* mRNAs. Further work will indicate their relevance in the regulation of mucin expression in the different parasite stages.

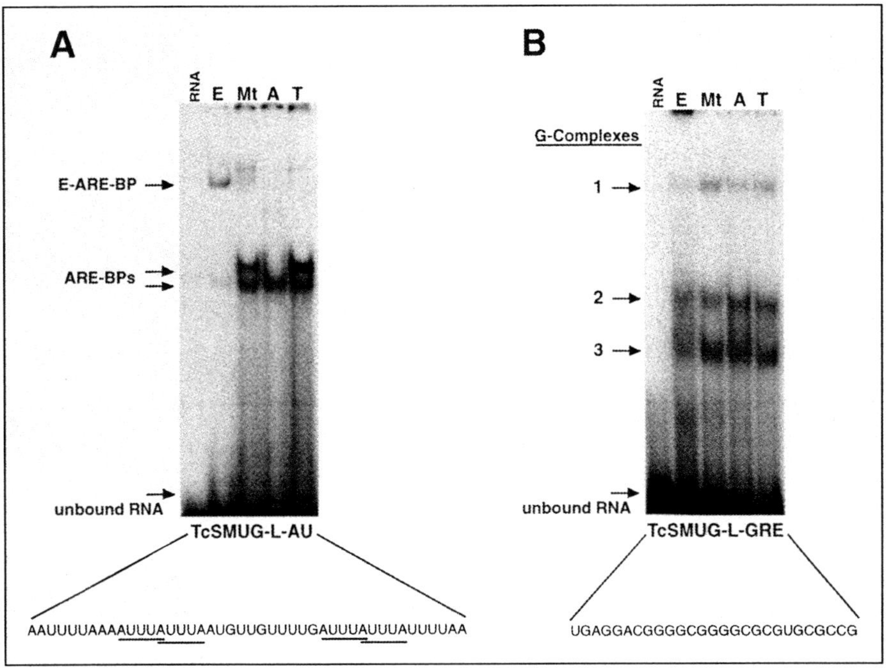

Figure 12. AU- and G-rich *cis*-elements form part of different ribonucleoprotein complexes. *Tcsmug*-L-AU and *Tcsmug*-L-GRE RNAs were incubated with total protein extract from the four lifecycle parasite stages: E, epimastigotes; Mt, metacyclic trypomastigotes; T, cell-derived trypomastigotes and A, intracellular amastigotes. The ribonucleoprotein complexes were separated by electrophoresis mobility shift assay (EMSA) on native polyacrylamide gels.

ARE-BPs and GRE-BPs As Part of a Model for Tcsmug *mRNA Stabilization*

A model for the post-transcriptional regulatory mechanism acting on *Tcsmug* mRNA and mediated by ARE and GRE RNA-binding proteins is shown in Figure 13. Both, the ARE deletion affecting *Tcsmug* mucin mRNA stability and the developmentally regulated expression pattern of the RNA-binding proteins that recognize the ARE motif, point to a coordinated and stage-specific process during the parasite lifecycle. E-ARE-BP, only expressed in the epimastigote stage, might be a positive *trans*-acting factor interacting with the ARE and protecting *Tcsmug* mRNA from degradation. E-ARE-BP binding could also prevent the association of destabilizing factor(s) to those mRNAs, possibly through competition for binding to similar *cis*-elements. Indeed, E-ARE-BP might be one of the proteins involved in the modulation of the translation activity mediated by the ARE motif,[35] probably through the interaction with other cellular factors of the translational apparatus.[72] On the other hand, GRE RNA-binding proteins are always present during the lifecycle of *T. cruzi* (Fig. 12). The possibility that an ARE-GRE-complex exists in vivo, and that this whole complex or some complex-forming proteins interact with a poly(A) binding protein or other cellular factor(s) to prevent the attack of a deadenylase activity in the polysome fraction, remains to be investigated. It is well known that in mammalian cells, a large complex is formed by several proteins that have different affinities for poly(C) homoribopolymer. An example is the assembly of the α-globin mRNA stability complex in the pyrimidine-rich region of the globin 3'UTR.[75]

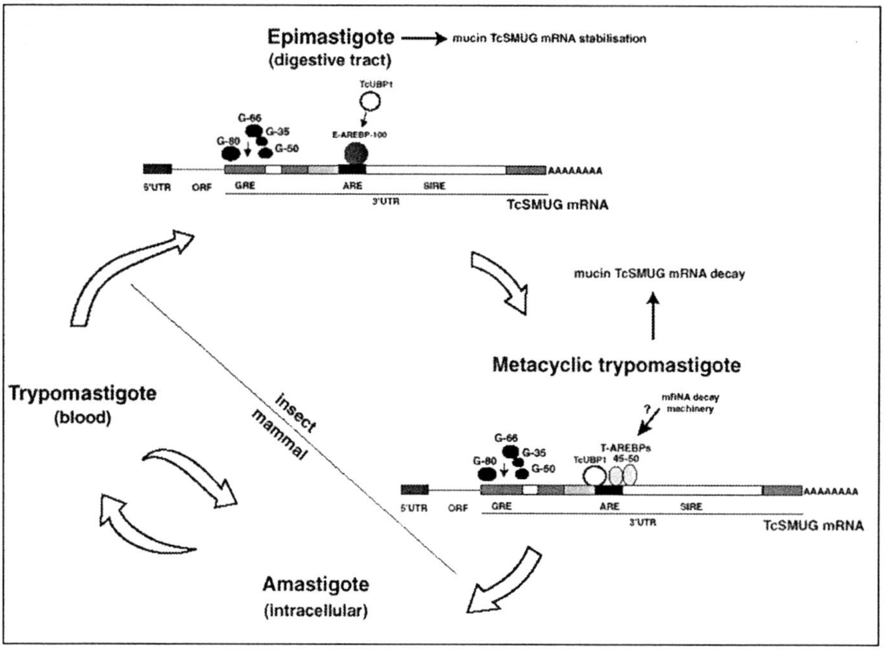

Figure 13. A model for the post-transcriptional regulatory mechanism acting on *Tcsmug* mRNA stability in different stages of *T. cruzi* development. GRE RNA-binding proteins (G-80, G-35, G-50 and G-66) are present in both epimastigote and metacyclic-trypomastigote stages of the parasite. E-ARE-BP (a 100 kDa ARE-binding protein) is only detected in the epimastigote stage and might protect mucin *Tcsmug* mRNA from degradation. The 45-50 kDa T-ARE-BPs are present in the other three developmental stages of the parasite. Additional cellular factors forming part of an mRNA decay machinery or translational apparatus might interact with the T-ARE-BPs and E-ARE-BP, respectively. In the amastigote and cell-derived trypomastigote stages, the model resembles that one represented for the metacyclic trypomastigote stage.

Post-transcriptional regulatory mechanisms, such as those mediated by ARE or GRE sequences, might be required for a quick response to change the pattern of TcSMUG expression, triggering parasite adaptation to sudden changes in the environment. In this regard, expression of the correct surface mucin coat might be of central importance for parasite survival. Identification of an in vivo role for these ARE and GRE RNA-binding proteins in the mRNA stability of *T. cruzi* transcripts might allow us to propose a model of RNA metabolism and maturation in parasites that appear to be deficient in the regulation of RNA-polymerase II-mediated transcription.

Future Work

The isolation of a defined homogenous mucin population, encoded by a single gene and from a particular parasite stage, so that its function can be studied without the noise caused by the presence of several different mucin populations, remains a challenge for the future. It has now become clear that mucins from *T. cruzi* are diverse and may have different functions. The identification of mucin genes may allow for the overexpression and purification of a single mucin species. The amino acid composition of the mucins isolated from epimastigotes, metacyclic trypomastigotes and cell-derived trypomastigotes, is not coincident with the predicted amino acid composition of the two groups of genes so far identified as encoding mucins. So, it is clear that more genes remain to be isolated or effectively identified. Furthermore, even when *T. cruzi* mucins where found to be adhesion molecules,[18,31] only one report has characterized their binding to L-selectin.[33] This is a major unexplored field given the obvious similarity of *T. cruzi* mucins with sialomucins that act as ligands for selectins and siglecs.[4] In this context, putative roles in parasite extravasation can be envisaged.

Regarding the mucin or mucin-like genes described so far in *T. cruzi*, a very complex system has been unveiled that pose a number of new questions. Group I genes from *Tcmuc* family were identified as encoding mucins from trypomastigotes, with a HV region exposed to the environment. The hypothesis about the function of this hypervariability should be tested during infection. For group II genes, questions include the nature of the products of the non-repetitive genes and the role of the high level of diversity. Are all these genes expressed and if so, does the variable region form part of the mature product? If this is the case, do the HV regions share a common function or have they diverged sufficiently to acquire different functions?. Variability does not always imply neutral evolution. Similarly, in the case of the group III genes, although the product has been detected and structurally and antigenically characterized, the function of the molecule has not yet been addressed. The S group of *Tcsmug* family encodes mucins from epimastigotes whose involvement in the interaction of the parasite with the insect has been proposed, but again this has never been addressed experimentally. Finally, the deduced products from the group L of this family have a structure resembling mucins but the products have not been identified.

In the field of regulation of mucin gene expression, the use of these two families as a model for studying post-transcriptional regulatory mechanisms has led to the identification of *cis*- and *trans*-acting components involved in the modulation of mRNA stability and translation. The specificity of the *trans*-acting factors should be studied in depth to identify the different mechanisms affecting pre-mRNA metabolism and maturation. It might be possible that different ARE-binding proteins form distinct ribonucleoprotein complexes, depending on the *cis*-element present. While the ARE within *Tcsmug*-L group 3'UTR is recognized by E-AREBP, other similar AU-rich regions are not recognized in vitro by this RNA-binding protein (our unpublished results). Recently, an ARE sequence was identified in the group III gene from the *Tcmuc* family. However, the product is only expressed in trypomastigote stages. As the AREs do not have a conserved sequence, but do share a common structure and composition, this might lead to the identification of new *trans*-acting factors that mediate stability or translation modulation in other parasite stages. The identification of a gene family encoding U-rich RNA-binding proteins binding to *Tcsmug* 3'UTR *cis*-elements and their preliminary characterization in terms of binding features and specificities would allow the study of their in vivo effect on mucin gene expression throughout development and their role in the regulation of the *Tcsmug* mRNA maturation process.

In conclusion, the identification of the *Tcmuc* and *Tcsmug* gene families has shed considerable light on the composition of the mucins present in the different trypanosome stages. The main contribution of these studies has been to show the high complexity of the mucin family and their regulation of expression. Now with the use of genetic manipulation techniques it should be possible to study mucin function in *T. cruzi* at the level of specific gene products, instead of at the level of complex populations of molecules.

Acknowledgements

The work performed in our laboratory was supported by grants from the Agencia Nacional de Promoción Científica y Tecnológica and the Consejo Nacional de Investigaciones Científicas y Técnicas (CONICET), Argentina, The Swedish Agency for Research Cooperation with Developing Countries (SIDA-SAREC) and the UNDP/World Bank/WHO Special Programme for Research and Training in Tropical Diseases (TDR). The work of ACCF was partially supported by an International Research Scholar grant from the Howard Hughes Medical Institute.

References

1. Van Klinken BJW, Einerhand AWC, Buller HA et al. Strategic biochemical analysis of mucins. Anal Biochem 1998; 265:103-116.
2. Van Klinken BJW, Dekker J, Buller HA et al. Mucin gene structure and expression: Protection vs adhesion. Am J Physiol 1995; 269:g613-g627.
3. Lasky LA, Singer MS,Dowbenko D et al. An endothelial ligand for L-selectin is a novel mucin-like molecule. Cell 1992; 69(6):927-938.
4. Lasky LA. Selectin-carbohydrate interactions and the initiation of the inflammatory response. Annu Rev Biochem 1995; 64:113-139.
5. Van den Steen P, Rudd PM, Dwek RA et al. Concepts and principles of O-linked glycosylation. Crit Rev Biochem Mol Biolog 1998; 33(3):151-208.
6. Jentoft N. Why are proteins O-glycosylated? Trends Biochem Sci 1990; 15(8):291-294.
7. Khoo KH, Maizels RM, Page AP et al. Characterization of nematode glycoproteins: the major O-glycans of Toxocara excretory-secretory antigens are O-methylated trisaccharides. Glycobiology 1991; 1(2):163-171.
8. Zinecker CF, Striepen B, Tomavo S et al. The dense granule antigen, GRA2 of Toxoplasma gondii is a glycoprotein containing O-linked oligosaccharides. Mol Biochem Parasitol 1998; 97(1-2):241-246.
9. Barnes DA, Bonnin A, Huang JX et al. A novel multi-domain mucin-like glycoprotein of Cryptosporidium parvum mediates invasion. Mol Biochem Parasitol 1998; 96(1-2):93-110.
10. Ilg T, Stierhof YD, Craik D et al. Purification and structural characterization of a filamentous, mucin-like proteophosphoglycan secreted by Leishmania parasites. J Biol Chem 1996; 271(35):21583-21596.
11. Lischke A, Klein C, Stierhof YD et al. Isolation and characterization of glycosylphosphatidylinositol-anchored, mucin-like surface glycoproteins from bloodstream forms of the freshwater-fish parasite Trypanosoma carassii. Biochem J 2000; 345 Pt 3:693-700.
12. Schenkman S, Ferguson MA, Heise N et al. Mucin-like glycoproteins linked to the membrane by glycosylphosphatidylinositol anchor are the major acceptors of sialic acid in a reaction catalyzed by trans-sialidase in metacyclic forms of Trypanosoma cruzi. Mol Biochem Parasitol 1993; 59(2):293-303.
13. Almeida IC, Ferguson MA, Schenkman S et al. Lytic anti-alpha-galactosyl antibodies from patients with chronic Chagas disease recognize novel O-linked oligosaccharides on mucin-like glycosyl-phosphatidylinositol-anchored glycoproteins of Trypanosoma cruzi. Biochem J 1994; 304(Pt 3):793-802.
14. Almeida IC, Krautz GM, Krettli AU et al. Glycoconjugates of Trypanosoma cruzi: A 74 kD antigen of trypomastigotes specifically reacts with lytic anti-alpha-galactosyl antibodies from patients with chronic Chagas disease. J Clin Lab Anal 1993; 7(6):307-316.
15. Schenkman S, Jiang MS, Hart GW et al. A novel cell surface trans-sialidase of Trypanosoma cruzi generates a stage-specific epitope required for invasion of mammalian cells. Cell 1991; 65(7):1117-1125.
16. Previato JO et al. Analysis of O-linked N-acetylglucosamine-containing oligosaccharides from sialic acid-acceptor glycoproteins expressed by trypomastigote forms of Trypanosoma cruzi (Colombiana strain) [Abstract]. Glycobiology 2000; 10.
17. Pereira-Chioccola VL, Acosta-Serrano A, Correia de Almeida I et al. Mucin-like molecules form a negatively charged coat that protects Trypanosoma cruzi trypomastigotes from killing by human anti-alpha- galactosyl antibodies. J Cell Sci 2000; 113(Pt 7):1299-1307.
18. Schenkman S, Kurosaki T, Ravetch JV et al. Evidence for the participation of the Ssp-3 antigen in the invasion of nonphagocytic mammalian cells by Trypanosoma cruzi. J Exp Med 1992; 175(6):1635-1641.
19. Franchin G, Pereira-Chioccola VL, Schenkman S et al. Passive transfer of a monoclonal antibody specific for a sialic acid- dependent epitope on the surface of Trypanosoma cruzi trypomastigotes reduces infection in mice. Infect Immun 1997; 65(7):2548-2554.
20. Camargo MM, Almeida IC, Pereira ME et al. Glycosylphosphatidylinositol-anchored mucin-like glycoproteins isolated from Trypanosoma cruzi trypomastigotes initiate the synthesis of proinflammatory cytokines by macrophages. J Immunol 1997; 158(12):5890-5901.

21. Almeida IC, Camargo MM, Procopio DO et al. Highly purified glycosylphosphatidylinositols from Trypanosoma cruzi are potent proinflammatory agents. Embo J 2000; 19(7):1476-1485.
22. Camargo MM, Andrade AC, Almeida IC et al. Glycoconjugates isolated from Trypanosoma cruzi but not from Leishmania species membranes trigger nitric oxide synthesis as well as microbicidal activity in IFN-gamma-primed macrophages. J Immunol 1997; 159(12):6131-6139.
23. Acosta-Serrano A, Schenkman RP, Schenkman S. Sialic acid acceptors of different stages of Trypanosoma cruzi are mucin-like glycoproteins linked to the parasite membrane by GPI anchors. Braz J Med Biol Res 1994; 27(2):439-442.
24. Yoshida N, Mortara RA, Araguth MF et al. Metacyclic neutralizing effect of monoclonal antibody 10D8 directed to the 35- and 50-kilodalton surface glycoconjugates of Trypanosoma cruzi. Infect Immun 1989; 57(6):1663-1667.
25. Di Noia JM, Pollevick GD, Xavier MT et al. High diversity in mucin genes and mucin molecules in Trypanosoma cruzi. J Biol Chem 1996; 271(50):32078-32083.
26. Mortara RA, da Silva S, Araguth MF et al. Polymorphism of the 35- and 50-kilodalton surface glycoconjugates of Trypanosoma cruzi metacyclic trypomastigotes. Infect Immun 1992; 60(11):4673-4678.
27. Singh BN, Lucas JJ, Beach DH et al. Expression of a novel cell surface lipophosphoglycan-like glycoconjugate in Trypanosoma cruzi epimastigotes. J Biol Chem 1994; 269(35):21972-21982.
28. Previato JO, Jones C, Xavier MT et al. Structural characterization of the major glycosylphosphatidylinositol membrane-anchored glycoprotein from epimastigote forms of Trypanosoma cruzi Y-strain. J Biol Chem 1995; 270(13):7241-7250.
29. Previato JO, Jones C, Goncalves LP et al. O-glycosidically linked N-acetylglucosamine-bound oligosaccharides from glycoproteins of Trypanosoma cruzi. Biochem J 1994; 301(Pt 1):151-159.
30. Acosta-Serrano A, Schenkman S, Yoshida N et al. The lipid structure of the glycosylphosphatidylinositol-anchored mucin- like sialic acid acceptors of Trypanosoma cruzi changes during parasite differentiation from epimastigotes to infective metacyclic trypomastigote forms. J Biol Chem 1995; 270(45):27244-27253.
31. Ruiz RDC, Rigoni VL, Gonzalez J et al. The 35/50 kDa surface antigen of Trypanosoma cruzi metacyclic trypomastigotes, an adhesion molecule involved in host cell invasion. Parasite Immunol 1993; 15(2):121-125.
32. Ruiz RC, Favoreto S Jr, Dorta ML et al. Infectivity of Trypanosoma cruzi strains is associated with differential expression of surface glycoproteins with differential Ca2+ signalling activity. Biochem J 1998; 330(Pt 1):505-511.
33. de Diego J, Punzon C, Duarte M et al. Alteration of macrophage function by a Trypanosoma cruzi membrane mucin. J Immunol 1997; 159(10):4983-4999.
34. Di Noia JM, Sanchez DO, Frasch AC. The protozoan Trypanosoma cruzi has a family of genes resembling the mucin genes of mammalian cells. J Biol Chem 1995; 270(41):24146-24149.
35. Di Noia JM, D'Orso I, Sanchez DO et al. AU-rich elements in the 3'-untranslated region of a new mucin-type gene family of Trypanosoma cruzi confers mRNA instability and modulates translation efficiency. J Biol Chem 2000; 275(14):10218-10227.
36. Reyes MB, Pollevick GD, Frasch AC. An unusually small gene encoding a putative mucin-like glycoprotein in Trypanosoma cruzi. Gene 1994; 140(1):139-140.
37. Freitas-Junior LH, Briones MR, Schenkman S. Two distinct groups of mucin-like genes are differentially expressed in the developmental stages of Trypanosoma cruzi [In Process Citation]. Mol Biochem Parasitol 1998; 93(1):101-114.
38. Salazar NA, Mondragon A, Kelly JM. Mucin-like glycoprotein genes are closely linked to members of the trans-sialidase super-family at multiple sites in the Trypanosoma cruzi genome. Mol Biochem Parasitol 1996; 78(1-2):127-136.
39. Di Noia JM, D'Orso I, Aslund L et al. The Trypanosoma cruzi mucin family is transcribed from hundreds of genes having hypervariable regions. J Biol Chem 1998; 273(18):10843-50I
40. Aguero F, Verdun RE, Frasch AC et al. A random sequencing approach for the analysis of the Trypanosoma cruzi genome: general structure, large gene and repetitive DNA families, and gene discovery. Genome Res 2000; 10(12):1996-2005.

41. Pollevick GD, Di Noia JM, Salto ML et al. Trypanosoma cruzi surface mucins with exposed variant epitopes. J Biol Chem 2000; 275(36):27671-27680.
42. Ohta T. Multigene families and the evolution of complexity. J Mol Evol 1991; 33(1):34-41.
43. Frasca L, Del Porto P, Tuosto L et al. Hypervariable region 1 variants act as TCR antagonists for Hepatitis C virus-specific CD4+ T cells. J Immunol 1999; 163:650-658.
44. Plebanski M, Flanagan KL, Lee EAM et al. Interleukin 10-mediated immunosuppression by a variant CD4 T cell epitope of Plasmodium falciparum. Immunity 1999; 10:651-660.
45. Millar AE, Wleklinski-Lee M, Kahn SJ. The surface protein superfamily of Trypanosoma cruzi stimulates a polarized Th1 response that becomes anergic. J immunol 1999; 162(10):6092-6099.
46. Maquat LE, Carmichael GG. Quality Control of mRNA Function. Cell 2001; 104(2):173-176.
47. Minvielle-Sebastia L, Keller W. mRNA polyadenylation and its coupling to other RNA processing reactions and to transcription. Curr Opin Cell Biol 1999; 11(3):352-357.
48. Cramer P et al. Coupling of transcription with alternative splicing: RNA pol II promoters modulate SF2/ASF and 9G8 effects on an exonic splicing enhancer. Mol Cell 1999; 4(2):251-8.I
49. Grosset C, Chen C, Xu YN et al. A mechanism for translationally coupled mRNA turnover: Interaction between the poly(A) tail and a c-fos RNA coding determinant via a protein complex. Cell 2000; 103(1):29-40.
50. Shaw G, Kamen R. A conserved AU sequence from the 3' untranslated region of GM-CSF mRNA mediates selective mRNA degradation. Cell 1986; 46(5):659-67I
51. Caput D, Beutler B, Hartog K et al. Identification of a common nucleotide sequence in the 3'-untranslated region of mRNA molecules specifying inflammatory mediators. Proc Natl Acad Sci USA 1986; 83(6):1670-1674.
52. Chen CY, Del Gatto-Konczak F, Wu Z et al. Stabilization of interleukin-2 mRNA by the c-Jun NH2-terminal kinase pathway. Science 1998; 280(5371):1945-1949.
53. Stoecklin G, Hahn S, Moroni C. Functional hierarchy of AUUUA motifs in mediating rapid interleukin-3 mRNA decay. J Biol Chem 1994; 269(46):28591-28597.
54. Kishore R, Tebo JM, Kolosov M et al. Cutting edge: clustered AU-rich elements are the target of IL-10-mediated mRNA destabilization in mouse macrophages. J Immunol 1999; 162(5):2457-2461.
55. Chen CA, Shyu A. AU-rich elements: characterization and importance in mRNA degradation. TIBS 1995; 20:465-470.
56. Kontoyiannis D, Pasparakis M, Pizarro T et al. Impaired on/off regulation of TNF biosynthesis in mice lacking TNF AU-rich elements: implications for joint and gut-associated immunopathologies. Immunity 1999; 10(3):387-398.
57. Sokolowski M, Zhao C, Tan W et al. AU-rich mRNA instability elements on human papillomavirus type 1 late mRNAs and c-fos mRNAs interact with the same cellular factors. Oncogene 1997; 15(19):2303-2319.
58. Kim YJ, Baker BS. Isolation of RRM-type RNA-binding protein genes and the analysis of their relatedness by using a numerical approach. Mol Cell Biol 1993; 13(1):174-183.
59. Fan XC, Steitz JA. Overexpression of HuR, a nuclear-cytoplasmic shuttling protein, increases the in vivo stability of ARE-containing mRNAs. Embo J 1998; 17(12):3448-3460
60. Loflin P, Chen CY, Shyu AB. Unraveling a cytoplasmic role for hnRNP D in the in vivo mRNA destabilization directed by the AU-rich element. Genes Dev 1999; 13(14):1884-1897.
61. Pays E, Nolan DP. Expression and function of surface proteins in Trypanosoma brucei. Mol Biochem Parasitol 1998; 91(1):3-36.
62. Roditi I, Furger A, Ruepp S et al. Unravelling the procyclin coat of Trypanosoma brucei. Mol Biochem Parasitol 1998; 91(1):117-130.
63. Berberof M, Vanhamme L, Tebabi P et al. The 3'-terminal region of the mRNAs for VSG and procyclin can confer stage specificity to gene expression in Trypanosoma brucei. Embo J 1995; 14(12):2925-2934.
64. Teixeira SM, Kirchhoff LV, Donelson JE. Post-transcriptional elements regulating expression of mRNAs from the amastin/tuzin gene cluster of Trypanosoma cruzi. J Biol Chem 1995; 270(38):22586-22594.
65. Matthews KR, Tschudi C, Ullu E. A common pyrimidine-rich motif governs trans-splicing and polyadenylation of tubulin polycistronic pre-mRNA in trypanosomes. Genes Dev 1994; 8(4):491-501.

66. McAndrew M, Graham S, Hartmann C et al. Testing promoter activity in the trypanosome genome: isolation of a metacyclic-type VSG promoter, and unexpected insights into RNA polymerase II transcription. Exp Parasitol 1998; 90(1):65-76.
67. Luo H, Gilinger G, Mukherjee D et al. Transcription initiation at the TATA-less spliced leader RNA gene promoter requires at least two DNA-binding proteins and a tripartite architecture that includes an initiator element. J Biol Chem 1999; 274(45):31947-31954.
68. Furger A, Schurch N, Kurath U et al. Elements in the 3' untranslated region of procyclin mRNA regulate expression in insect forms of Trypanosoma brucei by modulating RNA stability and translation. Mol Cell Biol 1997; 17(8):4372-4380.
69. Nozaki T, Cross GA. Effects of 3' untranslated and intergenic regions on gene expression in Tryanosoma cruzi. Mol Biochem Parasitol 1995; 75(1):55-67.
70. Grafi G, Sela I, Galili G. Translational regulation of human beta interferon mRNA: Association of the 3' AU-rich sequence with the poly(A) tail reduces translation efficiency in vitro. Mol Cell Biol 1993; 13(6):3487-3493.
71. Curatola AM, Nadal MS, Schneider RJ. Rapid degradation of AU-rich element (ARE) mRNAs is activated by ribosome transit and blocked by secondary structure at any position 5' to the ARE. Mol Cell Biol 1995; 15(11):6331-6340.
72. D'Orso I, Frasch ACC. Functionally different AU- and G-rich cis-elements confer developmentally regulated mRNA stability in Trypanosoma cruzi by interaction with specific RNA-binding proteins. J Biol Chem 2001; 276(19): 15783-15793.
73. Shyu AB, Wilkinson MF. The double lives of shuttling mRNA binding proteins. Cell 2000; 102(2):135-138.
74. Marchetti MA, Tschudi C, Kwon H et al. Import of proteins into the trypanosome nucleus and their distribution at karyokinesis. J Cell Sci 2000; 113(Pt 5):899-906.
75. Wang Z, Day N, Trifillis P et al. An mRNA stability complex functions with poly(A)-binding protein to stabilize mRNA in vitro. Mol Cell Biol 1999; 19(7):4552-4560.
76. Kelly JM, Ward HM, Miles MA et al. A shuttle vector which facilitates the expression of transfected genes in Trypanosoma cruzi and Leishmania. Nucleic Acids Res 1992; 20(15):3963-3969.

Chapter 4

How Does *Trypanosoma cruzi* Survive the Toxic Effects of Reactive Oxygen Species?

Shane R. Wilkinson and John M. Kelly

Abstract

Biological molecules are subject to attack by reactive oxygen species (ROS) leading to membrane disruption, inactivation of essential enzymes, mutagenesis and damage to DNA repair machinery. In aerobic organisms these ROS are produced by a number of endogenous processes and extensive systems have evolved to combat their toxic effects. Intracellular pathogens such as *Trypanosoma cruzi* are also exposed to ROS that can be generated by other mechanisms, including interactions with the host immune system and drug metabolism. It has long been considered that *T. cruzi* is deficient in its ability to deal with ROS.[1-3] This, together with observations that some mechanisms of oxidative defense in trypanosomatids are distinct from those of the mammalian host has encouraged the view that components of this system could provide targets for drug design.[3-5] The search for improved drugs against *T. cruzi* infection has received new impetus with reports confirming that the chronic disease process is dependent on the continued presence of the parasite. Most notably, studies in mice have shown that parasitic infection of the heart is both necessary and sufficient for the induction of cardiac tissue damage.[6] These observations therefore indicate that drugs targeted directly at the parasite have potential to reduce disease severity and/or progression. Over the last 2-3 years there have been major advances in our understanding of ROS detoxification in *T. cruzi*, particularly the enzymatic basis of peroxide metabolism. It is timely therefore to review these new findings and to assess their implications for our understanding of the role of oxidative defense in the *T. cruzi* life-cycle and the possibility that this system may provide opportunities for chemotherapeutic intervention.

What Are Reactive Oxygen Species?

Under aerobic conditions molecular oxygen is usually reduced to H_2O via 4-electron transfer. However several enzymes and small redox active molecules can reduce oxygen via 1- or 2-electron transfers resulting in the production of ROS which include superoxide anion (O_2^-), hydrogen peroxide (H_2O_2) and hydroxyl radical ($HO^·$) (Fig. 1). In most situations protective mechanisms within a cell are able to maintain ROS at nontoxic levels. Only when this balance is disturbed, either by increased production of ROS or by inhibition/blockage of the detoxification pathways, does the cell enter a state of oxidative stress. Under normal conditions O_2^- is converted to H_2O_2 by dismutation, a reaction catalysed by superoxide dismutase (SOD). H_2O_2 is metabolized by catalases or peroxidases to produce H_2O and O_2. In biological systems the most damaging ROS is the short-lived $HO^·$ (half-life 10^{-9} seconds), which upon formation

Molecular Mechanisms in the Pathogenesis of Chagas Disease, edited by John M. Kelly.

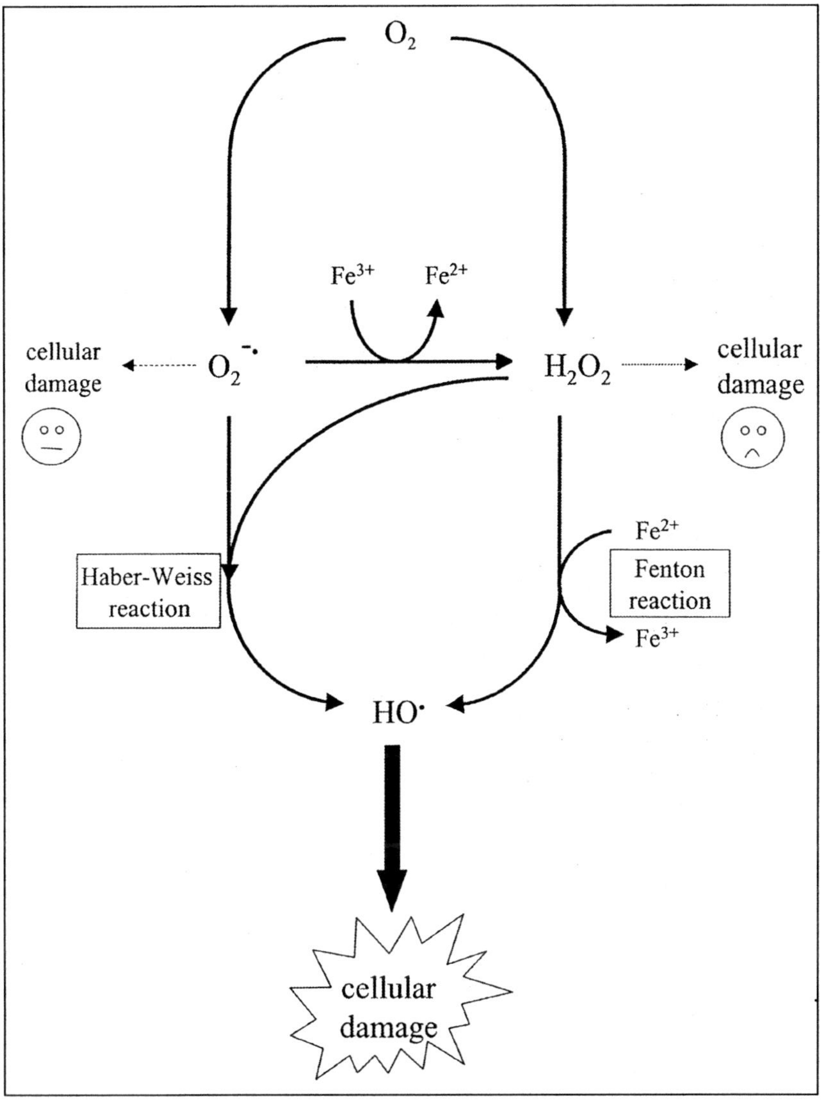

Figure 1. The nature of reactive oxygen species. Under certain conditions O_2 can undergo 1- or 2- electron reduction to generate reactive oxygen species such as $O_2^{\cdot-}$ or H_2O_2. These molecules, if left unchecked, may cause cellular damage either directly, by reacting with a variety of biological molecules, or indirectly via the action of highly reactive hydroxyl radicals ($HO^{\cdot}$) that are formed by the Fenton or Haber-Weiss reactions.

reacts immediately with the closest cellular component. $HO^{\cdot}$ can readily damage a number of biological molecules including nucleic acids, proteins and fatty acids. The latter are particularly susceptible to attack. This results in the formation of lipid peroxides ultimately leading to membrane destabilization.[7] $HO^{\cdot}$ can be formed from either O_2^{-} or H_2O_2 by the Fenton and Haber-Weiss reactions, processes that are catalysed by transition metals, usually copper or iron (Fig. 1). It is thought that the toxic effects caused by increased levels of H_2O_2 and O_2^{-} result mainly from the concomitant increase in $HO^{\cdot}$ formation. From this viewpoint, a major function of the antioxidant defense system is to prevent the build up of the ROS that lead to the production of HO·.

T. cruzi Is Exposed to Oxidative Stress Generated by Drug Metabolism and Immune Mechanisms

The nitroheterocyclic compounds nifurtimox and benznidazole have been the front line drugs used to treat acute stage *T. cruzi* infections.[8-10] However, both are considered to be unsatisfactory because of toxicity and limited efficacy. There is also little evidence to suggest that they are significantly effective against chronic Chagas disease,[11,12] the stage that has the greatest impact on public health. The precise mode of action of these drugs is unknown, although both have been demonstrated to undergo redox cycling within the parasite. Studies indicate that reduction of nifurtimox results in the generation of nitro-anion radicals followed by an autooxidation reaction that mediates the formation of toxic oxygen metabolites.[13,14] Although the enzymatic mechanisms responsible for this reduction have yet to be established in *T. cruzi*, it has been shown in the case of pig heart cells that dihydrolipoamide dehydrogenase is able to carry out this reaction with the resultant generation of O_2^-.[15] Functional analysis of the mitochondrially-targeted dihydrolipoamide dehydrogenase in *T. cruzi*[16] could shed further light on the mode of action of nifurtimox and may help to identify the cellular compartment in which the drug is metabolized. Benznidazole treatment also leads to the generation of O_2^- and H_2O_2, although to a lesser extent than nifurtimox.[17] Recent studies have indicated that in addition to redox cycling, both nifurtimox and benznidazole can exert their toxic effect by decreasing thiol levels within the parasite.[18] This is thought to result from conjugation of drug metabolites with glutathione and the parasite-specific thiol trypanothione (Fig. 2), thereby reducing the capacity of the cell to maintain a redox balance. The triarylmethane dye gentian violet has been used to prevent transmission of Chagas disease by blood transfusion.[19] Decontamination is effective but leads to staining of the blood and tissues. Oxidative stress resulting from the generation of O_2^- within the parasite has been implicated in the activity of this agent.[19,20] The effect of gentian violet is also enhanced by visible light by a mechanism thought to involve photochemical reduction of the drug to form a carbon-centred free radical.[21,22] This radical auto-oxidizes under aerobic conditions leading to the formation of O_2^-.

During an infection *T. cruzi* is also exposed to ROS that are generated as a result of the immune response. In γ-interferon (IFN-γ) activated macrophages the major trypanocidal activity involves the production of nitric oxide (NO) by inducible nitric oxide synthase.[23-25] In combination with O_2^-, this leads to the formation of peroxynitrite ($ONOO^-$), a potent biological oxidant that kills *T. cruzi* in a dose dependent manner.[26] The effects of $ONOO^-$ include thiol and lipid oxidation and nitrosylation of proteins. Inhibition of this trypanocidal response, by blocking either the production of NO or the transmission of the IFN-γ signal, has been shown to exacerbate infection. The main focus of the response in macrophages is directed toward the parasitophorous vacuole. The ability of the infective, trypomastigote form of *T. cruzi* to readily escape from this compartment into the cytoplasm therefore represents an effective strategy to avoid full exposure to the $ONOO^-$-mediated killing mechanism.

Thiol Metabolism in *T. cruzi* Is Unusual

All aerobic organisms have evolved a series of pathways that function to detoxify ROS before they cause serious biological damage. Collectively these form the oxidative defense system. Components of this system can be divided into either nonenzymatic antioxidants or enzyme-mediated pathways. In most eukaryotic cells the tripeptide glutathione (γ-glutamylcysteinylglycine; GSH) (Fig. 2 and 3) is the most important component of nonenzymatic oxidative defense. This low molecular weight thiol acts by maintaining an intracellular reducing

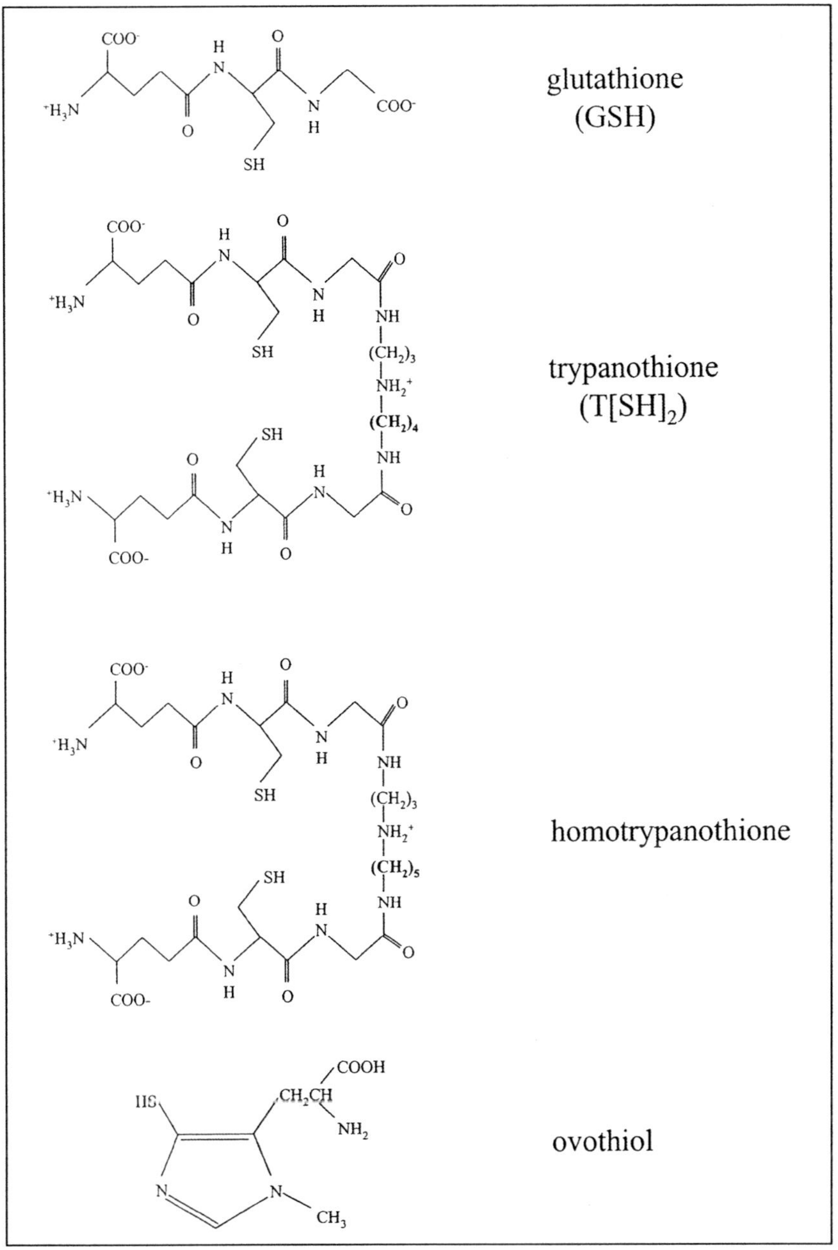

Figure 2. Structure of thiols present in trypanosomatids.

environment, has a major role as a free radical scavenger and is involved in a number of other cellular functions including DNA synthesis and detoxification of xenobiotics. GSH oxidation is also coupled to the reduction of H_2O_2 and lipid hydroperoxides by the activity of a number of GSH-dependent peroxidases (GPXs). GSH can then be regenerated from its oxidized form (GSSG) by the NADPH-dependent flavoenzyme glutathione reductase (GR).

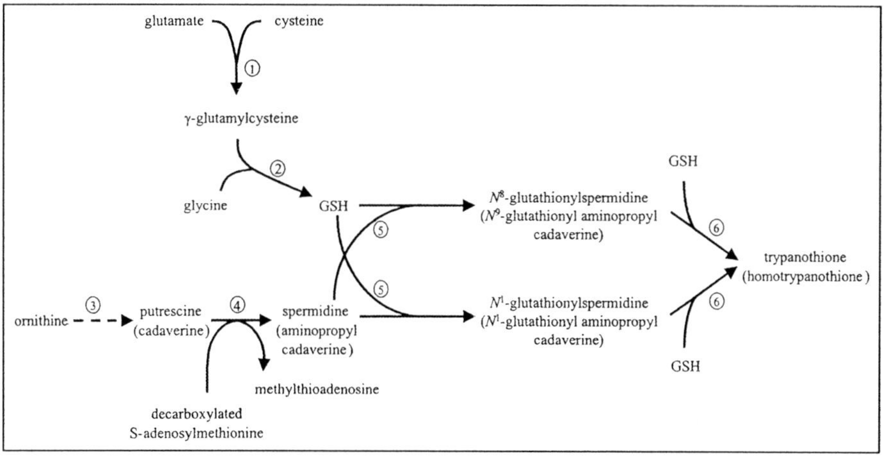

Figure 3. Trypanothione biosynthesis pathway. Trypanothione (or homotrypanothione) is synthesized from glutathione (GSH) and spermidine (or aminopropyl cadaverine) by the activity of glutathionylspermidine synthetase (5) and trypanothione synthetase (6). The GSH required for this synthesis is generated from glutamate, cysteine and glycine by the combined actions of γ-glutamylcysteine synthetase (1) and glutathione sythetase (2). In most trypanosomatids, spermidine is produced from ornithine, firstly by its decarboxylation to putrescine mediated by ornithine decarboxylase (3), followed by a nucleophilic substitution carried out by spermidine synthase (4). However, in *T. cruzi*, ornithine decarboxylase activity is reported to be absent. Instead the parasite scavenges the polyamines putrescine and cadaverine from the environment and these are then fed into the pathway as shown.

Thiol metabolism in trypanosomatids is distinct from that of other eukaryotes in two important ways.[4] They lack GR activity and the major low molecular mass thiol is not GSH, but typanothione ($T[SH]_2$), a GSH-spermidine conjugate (N^1,N^8-bis(glutathionyl)spermidine)[27] (Fig. 2). The synthesis of $T[SH]_2$ from GSH and spermidine is mediated by a unique pathway (Fig. 3) comprising two enzymes, glutathionylspermidine synthetase and trypanothione synthetase.[4,28,29] The parasite-specific nature of this pathway has led to the suggestion that it could represent a promising target for rational drug design.[3,30] GSH synthesis in trypanosomatids seems to occur by a pathway analogous to that in mammalian cells.[31] The first enzyme in the pathway, γ-glutamylcysteine synthetase, has been characterized in *Trypanosoma brucei*[32] and *Leishmania*[33] and evidence for the presence of the second, GSH synthetase, comes from the *Leishmania* Genome Project (Accession no. AL356246). In the absence of GR activity in trypanosomatids, GSH is maintained in its reduced form through an exchange reaction with $T[SH]_2$. In *T. cruzi* this can occur nonenzymatically or via a reaction catalysed by Tc52, a trypanothione-glutathione thioltransferase which shares motifs with glutathione-S-transferases of other organisms.[34,35] This protein, which is essential for viability,[36] has been implicated in intracellular growth and metacyclogenesis and has been proposed to have a role in the *T. cruzi*-mediated immunosuppression associated with Chagas disease.[37]

T. cruzi differs from other trypanosomatids such as *T. brucei* in that the major polyamine-thiol conjugate present in the cell is homotrypanothione (N^1,N^9-bis(glutathionyl)aminopropylcadaverine) (Fig. 2) rather than $T[SH]_2$.[38] This reflects a lack of ornithine decarboxylase activity in this parasite and a subsequent inability to synthesize putrescine[30,39] (there is also no evidence for a gene homologue in the *T. cruzi* database). Instead, diamines including putrescine and cadaverine, are scavenged from the environment[39,40] and *T. cruzi* has acquired the ability to use cadaverine as a precursor for polyamine synthesis. Homotrypanothione is formed by conjugation of aminopropylcadaverine and GSH in a reaction catalysed by trypanothione synthetase (Fig. 3). NADPH-dependent reduction of homotrypanothione disulfide by trypanothione reductase (TR) has been shown to be kinetically indistinguishable from the analogous reaction which uses trypanothione disulfide (T[S]$_2$).[38] Thus in *T. cruzi* it appears that homotrypanothione may be the main physiological substrate of this enzyme.

The flavoprotein oxidoreductase TR has an activity that is unique to the kinetoplastids and plays a central role in thiol metabolism (Fig. 4). Under normal circumstances this NADPH-dependent enzyme maintains more than 98% of the cellular trypanothione (or homotrypanothione) in its reduced form.[41] Over the last 10 years TR has attracted considerable attention as a possible chemotherapeutic target against Chagas disease and other trypanosomatid infections. As a consequence it has been subjected to detailed genetic, biochemical and physical analyses to validate it as a target and as an aid to "rational" drug design. The enzyme is a homodimer and has a reaction mechanism similar to that of human GR[4], with which it shares 40% amino acid sequence identity. However, the catalytic properties have been found to be highly specific in terms of substrate[42] and TR has no significant ability to reduce GSSG. This implies that it should be feasible to design inhibitors that can differentiate between TR and human GR, with the possibility that these compounds could function as trypanocidal agents. As an example of this, it has been reported that tricyclic compounds including acridines and phenothiazines, which are capable of competitive inhibition of TR but not GR,[43,44] have considerable activity against *T. cruzi* amastigotes. Likewise (terpyridine)platinum complexes, which can irreversibly inhibit TR but not GR, have cytostatic activity against trypanosomes.[45] An alternative chemotherapeutic strategy against *T. cruzi* has also been devised based on the substrate specificity of TR. This involves the use of "subversive" substrates such as nitrofuran derivatives.[46,47] Unlike GR, TR has the ability to catalyse the one–electron reduction of these trypanocidal agents. *In vivo*, this results in the generation of oxidative stress since the reduced product can undergo redox cycling leading to the production of O_2^-.

Two research groups have now resolved the three dimensional structure of *T. cruzi* TR using X-ray crystallography.[48-50] Further structural studies carried out on TR complexed with trypanothione[51] and enzyme inhibitors[52] have also provided greater insights into the precise catalytic mechanism and have provided a framework to facilitate the rational design of inhibitors.

TR is encoded by a single copy gene in pathogenic trypanosomatids.[53-55] It has therefore been possible to use targeted gene deletion and other reverse genetic approaches to investigate the biological function of the enzyme and to assess whether it represents a valid target for chemotherapy. In *Leishmania*, overexpression of TR does not confer an enhanced ability on the parasite to metabolize H_2O_2 or to survive the induction of oxidative stress.[56] This suggests that under these conditions the level of TR activity is not a limiting factor. However, when one of the chromosomal gene copies was deleted,[57,58] or 'dominant-negative' approaches were used to down-regulate TR activity,[59] *Leishmania* parasites exhibited a reduced ability to survive in activated macrophages. In addition, both chromosomal copies of the gene could only be deleted when cells had also been transformed with an episomal vector that expressed the TR gene. It can therefore be inferred that TR activity is essential for viability in the case of *Leishmania*. With *T. brucei*, null mutants devoid of TR activity could be isolated. However these cells were avirulent and exhibited increased sensitivity to oxidative stress.[60] In *T. cruzi*, TR gene knockout/disruption has yet to be reported, although attempts to down-regulate activity by expression of antisense RNA from a plasmid construct have been undertaken.[61] It was not possible in these experiments to produce genetically modified parasites that had reduced levels of TR. In all the transformants that were analysed, the input plasmid was found to have undergone rearrangements that prevented expression of the antisense transcript. This may indicate that down-regulation of TR activity is detrimental in *T. cruzi* and that under the culture conditions used in the experiment, only those cells that contained a rearranged plasmid could be selected.

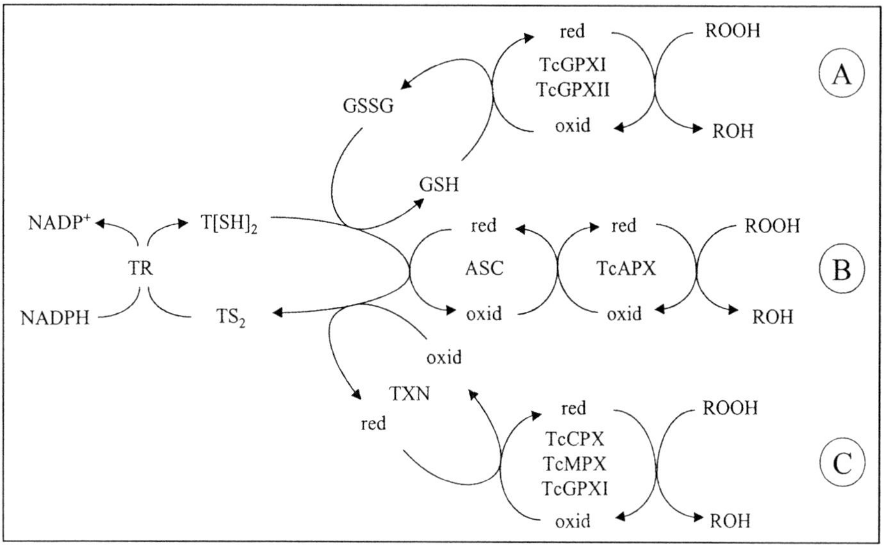

Figure 4. The trypanothione-dependent peroxide defense pathways identified within *T. cruzi*. Trypanothione disulphide (TS_2) is reduced to dihydrotrypanothione ($T[SH]_2$) by the NADPH-dependent flavoprotein trypanothione reductase (TR). There is now evidence for 3 distinct redox pathways linked to this reaction. (A) The glutathione peroxidase pathway: We have identified 2 enzymes with glutathione peroxidase activity (TcGPXI and TcGPXII). In this pathway, $T[SH]_2$ can interact with oxidized glutathione (GSSG) via nonenzymatic and enzymatic mechanisms to generate reduced glutathione (GSH). This then acts as an electron donor for TcGPXI and TcGPXII and facilitates the reduction of hydroperoxides (ROOH) to the corresponding alcohol (ROH). (B) The ascorbate-dependent pathway: In *T. cruzi,* ascorbate (ASC) is maintained in its reduced form by interaction with $T[SH]_2$ via a nonenzymatic mechanism. The enzyme ascorbate peroxidase (TcAPX) can then reduce hydroperoxides at the expense of ascorbate. This pathway should be regarded as putative, since it has yet to be reconstituted *in vitro*. (C) The tryparedoxin-mediated pathway: We have identified distinct mitochondrial (TcMPX) and cytosolic (TcCPX) peroxiredoxins in *T. cruzi*. Evidence indicates that TcCPX can reduce hydroperoxides using $T[SH]_2$ as an electron donor in a redox cascade that involves the thioredoxin-like molecule tryparedoxin (TXN). TcMPX may be part of a similar mitochondrial-localized pathway, although the details of this remain to be elucidated. We have also shown that the GSH-dependent peroxidase TcGPXI is also able to utilize tryparedoxin as an electron donor.

Other Possible Nonenzymatic Oxidative Defense Mechanisms

The parasite cell surface is the first site of contact with exogenously generated ROS, such as those produced during the macrophage oxidative burst. In *Leishmania,* the most abundant molecules on the surface of promastigotes is a glycosylphosphatidylinositol (GPI)-anchored phosphosaccharide called lipophosphoglycan (LPG).[62] These molecules have been shown to be highly efficient scavengers of O_2^- and $HO^{\cdot}$,[63] a factor that may contribute to pathogenicity.[64] *T. cruzi* also express abundant glycosinositol phospholiplids (GIPLs) on their cell surface. However the GIPLs lack the long phosphosaccharide repeats present on *Leishmania* LPG,[62] and it is not yet known whether these glycolipid molecules are a significant factor in oxidative defense. The major surface glycoproteins on the surface of *T. cruzi* are a heterogenous family of mucin-like molecules.[65] These highly glycosylated molecules also form a considerable protective barrier which could have a role in protecting the cell membrane from exogenous ROS.

In *T. cruzi* the presence of an additional redox system has been proposed based on the identification of ascorbate (vitamin C) in both epimastigotes and trypomastigotes.[66] This antioxidant has been shown to be an efficient scavenger of free radicals including O_2^- and $HO^{\cdot}$.

Ascorbic acid is maintained in a reduced state by dehydroascorbate reductase, the activity of which has been detected in *T. cruzi.*[66] However this activity has also been attributed to a nonenzymatic interaction with $T[SH]_2$.[67] There is also evidence in *T. cruzi* for the presence of an ascorbate-dependent enzyme-mediated redox cycle for the detoxification of H_2O_2 (Fig. 4). This follows from the identification of an EST (Accession no. AI057674) closely related to plant ascorbate-dependent peroxidases (identity in the range 40-45%) and the detection of the corresponding activity[1] in parasite extracts. The details and functional significance of this pathway remain to be fully elucidated.

Heterocyclic thiols have also been suggested to have a role in oxidative defense in trypanosomatids. In *Leishmania, T. brucei, T. cruzi* and the insect trypanosomatid *Crithidia fasciculata* the mercaptohistidine ovothiol A (1-*N*-methyl-4-mercaptohistidine) (Fig. 2) has been detected.[68,69,70] There is no evidence for the presence of an ovothiol reductase and evidence has shown that ovothiol A is maintained in its reduced form by interaction with $T[SH]_2$.[70] Ovothiols have very strong antioxidant and free radical-scavenging potentials and it has been proposed that this thiol may serve to protect *Leishmania* against ROS released by activated macrophages and from $ONOO^-$ mediated damage. In *T. cruzi* however, ovothiol A constitutes less than 10% of the total thiol content of the cell.[70] This, together with the finding that ovothiol A is a less efficient nonenzymatic scavenger of H_2O_2 than $T[SH]_2$ would seem to indicate that it does not have a major role in defense against this peroxide.

Trypanosomatids are also devoid of xanthine oxidase and thus uric acid, another potent antioxidant, is absent. Similarly vitamin E and β-carotene, which have important antioxidant roles in other organisms, have not been detected in *T. cruzi.*[71]

Dismutation of the Superoxide Anion Is Mediated by Fe-SODs in *T. cruzi*

SODs have been found in virtually all aerobic organisms, where they catalyse the dismutation of O_2^- to form H_2O_2 and O_2 (Fig. 1). Three major classes of SOD have been described on the basis of their divalent cation cofactors; Fe, Mn or Cu/Zn.[72] In *T. cruzi,* SOD activity has been shown to be cyanide insensitive/H_2O_2 and azide sensitive,[73,74] properties characteristic of the Fe-SOD isoform, the type usually found in prokaryotes and some plants. Typically these enzymes have monomeric sizes in the range 20-24 kDa and contain a single metal ion per monomer. In *T. cruzi* two *Fe-SOD* genes have been identified.[75,76] They appear to encode distinct mitochondrial and cytosolic variants, a pattern of subcellular distribution found in the majority of eukaryotes. O_2^- being a charged entity does not readily cross biological membranes and dismutation must occur within the compartment in which the radical is produced. Fe-SODs are characterized by presence of four amino acids that are localized in, or in proximity to, the active site.[77] These are present in both the *T. cruzi* enzymes (Ala-71, Gln-72, Tyr-79, Ala-145 in the cytosolic isoform).[75]

To investigate the role of SOD activity in *T. cruzi* under conditions of oxidative stress, epimastigotes were transfected with an expression vector containing the cytosolic *Fe-SOD* gene.[74] It was possible to achieve SOD activity in these cells that was up to 8 times higher than the normal level. Phenotypic analysis of the transformed parasites unexpectedly revealed that they were more susceptible than control cells to growth inhibition by benznidazole and gentian violet. A possible explanation for these findings is that the drug-mediated generation of O_2^- in cells that overexpress SOD could have caused an imbalance in the oxidative defense system due to the resulting increased rate of H_2O_2 production. In a situation where the ability to metabolize H_2O_2 is limited this could enhance the trypanocidal effect, either directly, or through the generation of $HO^·$. Indeed SOD itself has also been shown to catalyse $HO^·$ production from H_2O_2, and it has been suggested that this may explain, at least in part, the enhanced sensitivity to O_2^- of *E. coli* which overexpress Fe-SOD.[78] Both benznidazole and gentian violet undergo redox cycling in *T. cruzi,* with the concomitant production of O_2^-, but at a slow rate.[17,21,79]

Under normal conditions this mechanism is not considered to make a major contribution towards toxicity. However as has been observed with mammalian cells,[80] small changes to the physiological peroxidase/SOD activity ratios could have a major effect on the susceptibility of *T. cruzi* to oxidant damage.

In contrast to gentian violet and benznidazole, there was no increase in the sensitivity of cells overexpressing Fe-SOD to nifurtimox, nitrofurazone or menadione, agents that have also been shown to result in the generation of O_2^- in *T. cruzi*. The absence of conferred susceptibility may be a direct consequence of the subcellular compartment in which the redox-cycling occurs and the fact that when O_2^- is formed, it cannot subsequently cross lipid membranes. Although the enzymes that catalyse nifurtimox reduction *in vivo* have not been identified, redox-cycling of nifurtimox has been detected in *T. cruzi* mitochondrial fractions and it has been suggested that mitochondrial oxidoreductases such as dihydrolipoamide dehydrogenase[16] and cytochrome *c* reductase[79] may be involved. In this context, O_2^- generated by reductive metabolism of these drugs would be largely inaccessible to a cytosolic SOD. It should now be possible to further investigate these mechanisms using genetic techniques.

The Thioredoxin-Like Proteins

All the major trypanosomatid pathogens have been shown to express members of the thioredoxin family of oxidoreductases. These small proteins have been identified in a wide range of organisms where they function as electron donors in a number of cellular processes including ribonucleotide reduction, protein folding, peroxide metabolism, dehydroascorbate reduction and free radical scavenging.[81-83] A number of different types of thioredoxin molecules have been described. The different classes show little similarity to each other at the sequence level (<20%), but have a highly conserved 3-dimensional structure.[84] The only region with significant amino acid similarity across the group is at the dithiol active site, typically WCXXC. The two cysteines within this region mediate the thioredoxin redox activity by the formation/reduction of an intermolecular disulfide bond. To date three distinct families of thioredoxins have been identified in the trypanosomatids. A classical thioredoxin (active site, WCGPC) has been characterized from *T. brucei*[85] where it functions as an electron donor for ribonucleotide reductase, a role that can also be performed by $T[SH]_2$.[86] It has yet to be established if this thioredoxin is of importance in the context of oxidative defense. *C. fasciculata*,[87,88] *Leishmania major* (Accession nos. AQ849312, AQ901733) and *T. cruzi* (Accession nos. AAF04973, AI110351) have all been shown to contain a second distinct class of thioredoxin-like molecule, which have been called the tryparedoxins (active site, WCPPC).[89] The tryparedoxins act as electron donors to peroxiredoxins and form part of the cascade of oxidoreductases that mediate $T[SH]_2$-dependent peroxide metabolism (below, Fig. 4). A third thioredoxin, a member of the thioredoxin H family (WCGHC), has also been identified from a partial DNA sequence generated by the *T. cruzi* Genome Project (Accession no. AI069833). The role that this molecule plays within the parasite has yet to be determined. In other organisms thioredoxins are maintained in their reduced, active state by thioredoxin reductase, but to date no such enzyme activity has been reported in the trypanosomatids. In the case of the tryparedoxins, the reduced state results from interaction with $T[SH]_2$ (Fig. 4). Whether this is also the case with the other classes of thioredoxin-like molecules in *T. cruzi* is currently unknown.

The Trypanothione-Dependent Peroxiredoxin Pathway: A Central Player in Peroxide Metabolism

In most eukaryotes, catalases and GSH-dependent peroxidases are the front line enzymes in peroxide metabolism. Both types of enzyme have been reported to be absent from trypanosomes, although GPX activity has now been identified.[90] Earlier studies did provide biochemical evidence for $T[SH]_2$-mediated peroxidase activity in *T. brucei*, *Leishmania* and *Crithidia*.[91,92] However, until recently it was unclear how *T. cruzi* metabolized peroxides, and it had been proposed that this occurred by spontaneous reaction with $T[SH]_2$.[2] The situation was clarified following the identification of a $T[SH]_2$-dependent peroxide-metabolising redox pathway in the cytosol of *C. fasciculata*.[87,89] This pathway (Fig. 4) utilizes the NADPH-dependent reduction of $T[S]_2$ by TR to drive a two-step oxidoreductase cascade involving tryparedoxin and tryparedoxin peroxidase, a member of the peroxiredoxin family of antioxidant proteins. The tryparedoxin peroxidase has two domains that are characteristic of the 2-cys subgroup of the peroxiredoxin family, which contain the cysteine residues that have been implicated in mediating peroxidase activity.[93,94] Structural studies by X-ray crystallography support this mechanism for the *Crithidia* enzyme.[95]

In *T. cruzi* two distinct peroxiredoxins with trypanothione-dependent peroxidase activity have now been reported.[96] One is cytosolic (TcCPX) and the other (TcMPX) has been localized to the mitochondrion. The enzymes are able to reduce both H_2O_2 and small organic peroxides such as *t*-butylhydroperoxide. The activity of TcCPX is mediated by tryparedoxin[97] (Fig. 4), and its widespread distribution throughout the cytosol[96] suggests a general peroxide-scavenging role. With TcMPX, the tryparedoxin linkage has yet to be confirmed. However the mitochondrial location of the enzyme implies that either the mitochondrion contains all the redox machinery required to maintain the enzyme in its reduced form, or the redox pathway is partitioned between the cytosol and the mitochondrion. The subcellular distribution of $T[SH]_2$ has yet to be reported, but biochemical evidence from *T. brucei* strongly suggests that the ancillary enzyme TR is restricted to the cytosol.[98] If this is also the case with *T. cruzi*, the identification of a mitochondrial enzyme that is $T[SH]_2$-dependent suggests the presence of an unidentified thiol transporter and/or reduction system. Such a system would be of considerable interest, particularly as a target for chemotherapy. In other eukaryotic cells, GSH is present at high levels in the mitochondria. In the absence of a biosynthetic pathway within the organelle, GSH is translocated from the cytosolic pool to the mitochondrial matrix. The exact transport mechanism is unknown but may be multicomponent[99] and involve the dicarboxylate and 2-oxoglutarate carriers of the mitochondrial inner membrane.[100] The presence of other GSH transporter systems has not been excluded. In *T. cruzi*, because of the distinct structure of $T[SH]_2$, it is likely that any transporter/exchanger system will differ from that in the mammalian host. It is also possible that TcMPX is part of a mitochondrial redox pathway, analogous to that present in the cytosol. Mitochondrial-localized thioredoxins have been identified in mammalian and yeast cells,[82,101,102] but not so far in trypanosomes. The precise biological role of the mitochondrial peroxidase remains to be determined. The observed concentration of TcMPX in the vicinity of the kinetoplast[96] suggests that a major function of this enzyme may be to protect the mitochondrial genome from direct, or indirect peroxide-mediated damage.

Parasites genetically modified to overexpress TcMPX or TcCPX were found to have increased resistance to H_2O_2 and *t*-butylhydroperoxide.[96] Peroxides are uncharged molecules and readily cross plasma or organelle membranes. The finding that overexpression of compartmentalized peroxidases in *T. cruzi* protects the cell from the toxic effects of exogenous peroxides indicates that these oxidants gain entry to the cytosol and the mitochondrion and that increased peroxide metabolism at these sites can confer protection against cellular damage. Thus in their biological context, both TcMPX and TcCPX are capable of protecting the parasite from peroxides of both endogenous and exogenous origin. Previous experiments with *Leishmania* have shown that the level of TR activity is not a rate-limiting step in peroxide metabolism in situations where this enzyme is overexpressed.[56] The observation that overexpression of peroxidase activity in transformed cells is associated with enhanced resistance implies that the level of both TcCPX and TCMPX is a rate-limiting factor in peroxide metabolism within their respective subcellular compartments. These transformed cells were also tested to determine if they had acquired increased resistance to benznidazole or nifurtimox. No differences were observed suggesting that the predominant killing activity of these drugs is not by peroxide-mediated damage.

The Glutathione-Dependent Peroxidase Pathways: An Unexpected Discovery

The rapid progress of the *T. cruzi* Genome Project is now having a major impact on many aspects of research, including the unravelling of the parasite oxidative defense mechanisms. As an example of this, we have identified sequences from the *T. cruzi* EST database (Accession nos. AI069627, AI035005 and AI046216) that appear to correspond to two distinct GSH-dependent peroxidases. This was an unexpected observation given that many studies have reported that trypanosomatids lack this activity (for review, ref. 4). The two enzymes, which we have now fully sequenced, have been designated TcGPXI[90] and TcGPXII (manuscript in preparation). They have highest similarity to plant GPXs and mammalian phospholipid hydroperoxide glutathione peroxidases (PHGPX) which together form a discrete clade (the PHGPX clade) within the GPX family.[103] When compared to representatives from the other GPX clades, the PHGPX proteins lack the two sequence insertions that are thought to mediate oligomerization.[103] This may explain the monomeric nature of PHGPX proteins. It is also believed that these insert regions could play a role in determining substrate specificity; PHGPXs can metabolize phospholipid hydroperoxides, whereas some other GPXs cannot. As well as these differences, mammalian PHGPXs are also distinct from *T. cruzi* and plant GPXs in that the crucial redox active amino acid in their active sites is a selenocysteine. In contrast the corresponding residue in the trypanosomal[90] and the plant GPXs[104,105] is a cysteine. This residue, in conjunction with two others, forms the catalytic triad that mediates the peroxide reduction activity. The interchange of selenocysteine/cysteine at this site has been shown to play an important role in antioxidant function by affecting both peroxidase activity and sensitivity towards peroxide-mediated inactivation.[104,105]

To confirm that sequence extends to function, we have now shown that recombinant versions of both TcGPXI and TcGPXII have the ability to reduce peroxides in the presence of GSH/GR, but not in the presence of $T[SH]_2$/TR.[90] Differences in substrate specificity have been observed between the two enzymes. TcGPXI can metabolize a wide range of hydroperoxides (linoleic acid hydroperoxide and phosphatidylcholine hydroperoxide > cumene hydroperoxide > *t*-butylhydroperoxide), but no activity towards H_2O_2 has been detected.[90] In contrast, the peroxidase activity of TcGPXII has only been detected when using linoleic acid hydroperoxide or phosphatidylcholine hydroperoxide as substrate (manuscript in preparation). Preliminary studies using antisera raised against both TcGPXI and TcGPXII suggest that both enzymes are associated with membranes. Therefore, on the basis of substrate specificity and the localization data, the main role of these enzymes in *T. cruzi* may be to minimize or prevent cellular damage due to lipid peroxidation. To gain further insight into the function of these peroxidases within the parasite, recombinant cell lines were generated that overexpressed either TcGPXI or TcGPXII. These cells were found to possess increased resistance towards exogenous H_2O_2, even though neither enzyme is able to metabolize this peroxide. Our interpretation of these results is that TcGPXI and TcGPXII have a detoxification role in *T. cruzi* and that they reduce the lipid peroxides that arise from the oxidising activity of H_2O_2.

Like many other GPXs, both TcGPXI and TcGPXII lack some of the residues that have been demonstrated to contribute to GSH binding in the cytosolic GPXs.[104] This has led to the suggestion that GSH may not be the only physiological reductant for these enzymes. To investigate if this was the case for TcGPXI and TcGPXII, assays were carried out in the presence of $T[SH]_2$/TR and dialysed *T. cruzi* extract. With TcGPXI, but not TcGPXII, a factor in the lysate was found to have the ability to link peroxidase activity to the reduction of $T[SH]_2$ (Fig. 4).

Investigation into the nature of this factor resulted in the purification of a tryparedoxin molecule that could reconstitute this redox cascade *in vitro* (manuscript in preparation). GPXs with thioredoxin-dependent activity have been reported previously and are generally localized in areas where the concentration of GSH is low, for example within the blood plasma.[106] Thus the ability of TcGPXI to scavenge reducing equivalents from different sources may reflect that one of the components of either the tryparedoxin or GSH redox cycle is present at a low level in the cellular environment where the peroxidase is found.

Summary

The biochemical basis of oxidative defense in *T. cruzi* is rapidly being dissected. Advances in this area have been greatly facilitated by the Trypanosomatid Genome Projects. It is now apparent that the enzymatic arm of this defense system is more complex than previously realized, and that a number of overlapping pathways are involved (Fig. 4). This has implications for our understanding of how *T. cruzi* survives the toxic effects of ROS, including those that can be generated by immune mechanisms and drug metabolism. With the application of genetic manipulation procedures and new "functional genomic" approaches, it should now be feasible to isolate and functionally characterize the entire repertoire of oxidative defense enzymes. An outcome of this work should be the identification and validation of several new targets that have potential for the design of drugs active against *T. cruzi.*

Acknowledgements

We thank the Wellcome Trust and the British Heart Foundation for financial support, and Martin Taylor and David Meyer for constructive comments on the preliminary draft this chapter.

References

1. Boveris A, Sies H, Martino EE et al. Deficient metabolic utilization of hydrogen peroxide in Trypanosoma cruzi. Biochem J 1980; 188:643-648.
2. Carnieri EVS, Moreno SNJ, Docampo R. Trypanothione-dependent peroxide metabolism in Trypanosoma cruzi different stages. Mol Biochem Parasitol 1993; 61:79-86.
3. Krauth-Siegel RL, Coombs GH. Enzymes of parasite thiol metabolism as drug targets. Parasitol Today 1999; 15:404-409.
4. Fairlamb AH, Cerami A. Metabolism and functions of trypanothione in the kinetoplastida. Annu Rev Microbiol 1992; 46:695-729.
5. Flohe L, Hecht HJ, Steinert P. Glutathione and trypanothione in parasitic hydroperoxide metabolism. Free Radic Biol Med 1999; 27:966-984.
6. Tarleton RL, Zhang L, Downs MO. "Autoimmune rejection" of neonatal heart transplants in experimental Chagas disease is a parasite-specific response to infected tissue. Proc Natl Acad Sci USA 1997; 94:3932-3937.
7. Acworth IN, Bailey B. The Handbook of Oxidative Metabolism. Chelmsford: ESA Inc., 1995.
8. Gustafsson LL, Beerman B, Abdi YA. Handbook For Drugs For Tropical Diseases. London: Taylor and Francis Ltd., 1987.
9. Van Gompel A, Vervoort T. Chemotherapy of leishmaniasis and trypanosomiasis. Curr Opin Infect Dis 1997; 10:469-474.
10. Urbina JA. Chemotherapy of Chagas disease: the how and the why. J Mol Med 1999; 77:332-338.
11. de Castro SL. The challenge of Chagas disease chemotherapy: an update of drugs assayed against Trypanosoma cruzi. Acta Trop 1993; 53:83-98.
12. Viotti R, Vigliano C, Armenti H et al. Treatment of chronic Chagas disease with benznidazole: clinical and serologic evolution of patients with long-term follow-up. Am Heart J 1994; 127:151-162.
13. Docampo R, Moreno SNJ. Free Radicals in Biology. Pryor WA, ed. Vol. VI. New York: Academic Press, 1984:243-288.

14. Docampo R. Sensitivity of parasites to free radical damage by antiparasitic drugs. Comp Biol Interactions 1990; 73:1-27.
15. Sreider CM, Grinblat L, Stoppani AOM. Reduction of nitrofuran compounds by heart lipoamide dehydrogenase : role of flavin and the reactive disulfide groups. Biochem Int 1992; 28:323-334.
16. Schoneck R, Billaut-Mulot O, Numrich P. et al. Cloning, sequencing and functional expression of dihydrolipoamide dehydrogenase from the human pathogen Trypanosoma cruzi. Eur J Biochem 1997; 243:739-747.
17. Masana M, Toranzo EGD, Castro AC. Reductive metabolism and activation of benznidazole. Biochem Pharmacol 1984; 33:1041-1045.
18. Maya JD, Repetto Y, Agosin M et al. Effects of nifurtimox and benznidazole upon glutathione and trypanothione content in epimastigote, trypomastigote and amastigote forms of Trypanosoma cruzi. Mol Biochem Parasitol 1997; 86:101-106.
19. Docampo R, Moreno SNJ, Gadelho FR et al. Prevention of Chagas disease resulting from blood transfusion by treatment of blood: toxicity and mode of action of gentian violet. Biomed Environ Sciences 1988; 1:406-413.
20. Docampo R, Moreno SNJ, Cruz FS. Enhancement of the cytotoxicity of crystal violet against Trypanosoma cruzi by ascorbate. Mol Biochem Parasitol 1988; 27:241-248.
21. Docampo R, Moreno SNJ, Muniz RPA et al. Light-enhanced free radical formation and trypanocidal action of gentian violet (crystal violet). Science 1983; 220:1292-1295.
22. Reske K, Cruz FS, Docampo R. Photosensitization by the trypanocidal agent crystal violet. Type I versus type II reactions. Chem Biol Interactions 1986; 58:161-172.
23. Vespa GN, Cunha FQ, Silva JS. Nitric oxide is involved in control of Trypanosoma cruzi-induced parasitemia and directly kills the parasite in vitro. Infect Immun 1994; 62:5177-5182.
24. Norris KA, Schrimpf JE, Flynn JL et al. Enhancement of macrophage microbicidal activity: supplemental arginine and citrulline augment nitric oxide production in murine peritoneal macrophages and promote intracellular killing of Trypanosoma cruzi. Infect Immun 1995; 63:2793-2796.
25. Jacobs F, Chaussabel D, Truyens C et al. IL-10 up-regulates nitric oxide (NO) synthesis by lipopolysaccharide (LPS)-activated macrophages: improved control of Trypanosoma cruzi infections. Clin Exp Immunol 1998; 113:59-64.
26. Denicola A, Rubbo H, Rodriguez D et al. Peroxynitrite-mediated cytotoxicity to T. cruzi. Arch Biochem Biophys 1993; 304:279-287.
27. Fairlamb AH, Blackburn P, Ulrich P et al. Trypanothione: a novel bis (glutathionyl) sperimidine cofactor for glutathione reductase in trypanosomatids. Science 1985; 227:1485-1487.
28. Smith K, Nadeau K, Walsh CT et al. Purification of glutathionylspermidine and trypanothione synthetases from Crithidia fasciculata. Protein Sci 1992; 1:874-883.
29. Tetaud E, Manai F, Barrett MP et al. Cloning and characterization of the two enzymes responsible for trypanothione biosynthesis in Crithidia fasciculata. J Biol Chem 1998; 273:19383-19390.
30. Fairlamb AH. Future prospects for chemotherapy of Chagas disease. Medicina 1999; 59 (Supl II):179-180.
31. Arrick BA, Griffith OW, Cerami A. 1981 Inhibition of glutathione synthesis as a chemotherapeutic strategy for trypanosomiasis. J Exp Med 1981; 153:720-725.
32. Lueder DV, Phillips MA. Characterization of Trypanosoma brucei γ-glutamylcysteine synthetase, an essential enzyme in the biosynthesis of trypanothione (diglutathionylspermidine). J Biol Chem 1996; 271:17485-17490
33. Grondin K, Haimeur A, Mukhopadhyay R et al. Co-amplification of the γ-glutamylcysteine synthetase gene gsh1 and of the ABC transporter gene pgpA in arsenite–resistant Leishmania tarentolae. EMBO J 1997; 16:3057-3065.
34. Moutiez M, Aumercier P, Schoneck R et al. Purification and characterisation of a trypanothione-glutathione thioltransferase from Trypanosoma cruzi. Biochem J 1995; 310:433-437.
35. Moutiez M, Quemeneur E, Sergheraert C et al. Glutathione-dependent activities of Trypanosoma cruzi p52 makes it a member of the thiol:disulphide oxidoreductase family. Biochem J 1997; 322:43-48.
36. Allaoui A, Francois C, Zemzoumi K et al. Intracellular growth and metacyclogenesis defects in Trypanosoma cruzi carrrying a targeted deletion of a Tc52 protein-encoding allele. Mol Microbiol 1999; 32:1273-1286.

37. Fernandez-Gomez R, Esteban S, Gomez-Corvera R, Zoulika K, Ouaissi, A. Trypanosoma cruzi: Tc52 released protein-induced increased expression of nitric oxide synthase and nitric oxide production by macrophages. J Immunol 1998; 160:3471-3497.
38. Hunter KJ, Le Quesne SA, Fairlamb AH. Identification and biosynthesis of N^1,N^9-bis(glutathionyl)aminopropylcadaverine (homotrypanothione) in Trypanosoma cruzi. Eur J Biochem 1994; 226:1019-1027.
39. Ariyanaygam MR, Fairlamb AH. Diamine auxotrophy may be a universal feature of Trypanosoma cruzi epimastigotes. Mol Biochem Parasitol 1997; 84:111-121.
40. Le Quesne SA, Fairlamb AH. Regulation of a high affinity diamine transport system in Trypanosoma cruzi epimastigotes. Biochem J 1996; 316:481-486.
41. Fairlamb AH, Henderson GB, Bacchi CJ et al. In vivo effects of difluoromethylornithine on trypanothione and polyamine levels in Trypanosoma brucei. Mol Biochem Parasitol 1987; 24:185-191.
42. Marsh IR, Bradley M. Substrate specificity of trypanothione reductase. Eur J Biochem 1997; 243:690-694.
43. Krauth-Siegel RL, Schoneck R. Flavoprotein structure and mechanism. 5. Trypanothione reductase and lipoamide dehydrogense as targets for structure-based drug design. FASEB J 1995; 12:1138-1146.
44. Khan MO, Austin SE, Chan C et al. Use of an additional hydrophobic binding site, the Z site, in rational drug design of a new class of stronger trypanothione reductase inhibitor, quaternary alkylammonium phenothiazines. J Med Chem 2000; 43:3148-3156.
45. Bonse S, Richards JM, Ross SA et al. (2,2':6',2"-Terpyridine)platinum (II) complexes are irreversible inhibitors of Trypanosoma cruzi trypanothione reductase but not of human glutathione reductase. J Med Chem 2000; 43:4812-4812.
46. Henderson GB, Ulrich P, Fairlamb AH et al. "Subversive" substrates for the enzyme trypanothione reductase: alternative approaches to chemotherapy of Chagas disease. Proc Natl Acad Sci USA 1988; 85:5374-5378.
47. Blumenstiel K, Schoneck R, Yardley V et al. Nitrofuran drugs as common subversive substrates of Trypanosoma cruzi lipoamide dehydrogenase and trypanothione reductase. Biochem Pharmacol 1999; 58:1791-1799.
48. Zhang Y, Bailey S, Naismith JH et al. Trypanosoma cruzi trypanothione reductase. Crystallization, unit cell dimensions and structure solution. J Mol Biol 1993; 232:1217-1220.
49. Lantwin CB, Schlichting I, Kabsch W et al. The structure of Trypanosoma cruzi trypanothione reductase in the oxidised and NADPH reduced state. Proteins 1994; 18:161-173.
50. Zhang Y, Bond CS, Bailey S et al. The crystal structure of trypanothione reductase from the human pathogen Trypanosoma cruzi at 2.3 A resolution. Protein Sci 1996; 5:52-61.
51. Bond CS, Zhang Y, Berriman M et al. Crystal structure of Trypanosoma cruzi trypanothione reductase in complex with trypanothione, and the structure-based discovery of new natural product inhibitors. Structure Fold Des 1999; 7:81-89.
52. Jacoby EM, Schlichting I, Lantwin CB et al. Crystal structure of the Trypanosoma cruzi trypanothione reductase.mepacrine complex. Proteins 1996; 24:73-80.
53. Sullivan FX, Walsh CT. Cloning, sequencing, overproduction and purification of trypanothione reductase from Trypanosma cruzi. Mol Biochem Parasitol 1991; 44:145-147.
54. Aboagye-Kwarteng T, Smith K, Fairlamb AH. Molecular characterisation of the trypanothione reductase gene from Crithidia fasciculata and Trypanosoma brucei: comparison with other flavoprotein disulphide oxidoreductases with respect to substrate specificity and catalytic mechanism. Mol Microbiol 1992; 6:3089-3099.
55. Taylor MC, Kelly JM, Chapman CJ et al. The structure, organisation, and expression of the Leishmania donovani gene encoding trypanothione reductase. Mol Biochem Parasitol 1994; 64:293-301.
56. Kelly JM, Taylor MC, Smith K et al. Phenotype of recombinant Leishmania donovani and Trypanosoma cruzi which overexpress trypanothione reductase. Sensitivity towards agents that are thought to induce oxidative stress. Eur J Biochem 1993; 218:29-37.
57. Dumas C, Ouellette M, Tovar J et al. Disruption of the trypanothione reductase gene of Leishmania decreases its ability to survive oxidative stress in macrophages. EMBO J 1997; 15:2590-2598.
58. Tovar J, Wilkinson S, Mottram JC et al. Evidence that trypanothione reductase is an essential enzyme in Leishmania by targeted replacement of the tryA gene locus. Mol Microbiol 1998; 29:653-660.

59. Tovar J, Cunningham ML, Smith AC et al. Down-regulation of Leishmania donovani trypanothione reductase by heterologous expression of a trans-dominant mutant homologue: effect on parasite intracellular survival. Proc Natl Acad Sci USA 1998; 95:5311-5316.
60. Krieger S, Schwarz W, Ariyanayagam MR et al. Trypanosomes lacking trypanothione reductase are avirulent and show increased sensitivity to oxidative stress. Mol Microbiol 2000; 35:542-552.
61. Tovar J, Fairlamb AH. Extrachromosomal, homologous expression of trypanothione reductase and its complementary mRNA in Trypanosoma cruzi. Nucl Acid Res 1996; 24:2942-2949.
62. Ferguson MAJ. The structure, biosynthesis and functions of glycosylphosphatidylinositol anchors, and the contributions to trypanosome research. J Cell Sci 1999; 112:2799-2809.
63. Chan J, Fujiwara T, Brennan P et al. Microbial glycolipids: possible virulence factors that scavenge oxygen radicals. Proc Natl Acad Sci USA 1989; 86:2453-2457.
64. Turco SJ, Descoteaux A. The lipophosphoglycan of Leishmania parasites. Annu Rev Microbiol 1992; 46:65-94.
65. Frasch AC. Functional diversity in the trans-sialidase and mucin families in Trypanosoma cruzi. Parasitol Today 2000; 16:282-286.
66. Clark D, Albrecht M, Arevalo J. Ascorbate variations and dehydroascorbate reductase activity in Trypanosoma cruzi epimastigotes and trypomastigotes. Mol Biochem Parsitol 1994; 66:143-145.
67. Krauth-Siegel RL, Ludemann H. Reduction of dehydroascorbate by trypanothione. Mol Biochem Parasitol 1996; 80:203-208.
68. Steenkamp DJ, Spies HS. Identification of a major low-molecular-mass thiol of the trypanosomatid Crithidia fasciculata as ovothiol A. Facile isolation and structural analysis of the bimane derivative. Eur J Biochem 1994; 223:43-50.
69. Spies HS, Steenkamp DJ. Thiols of intracellular pathogens. Identification of ovothiol A in Leishmania donovani and structural analysis of a novel thiol from Mycobacterium bovis. Eur J Biochem 1994; 224:203-213.
70. Ariyanayagam MR, Fairlamb AH. Ovothiol and trypanothione as antioxidants in trypanosomatids. Mol Biochem Parasitol 2001; 115:189-198.
71. Docampo R. Antioxidant mechanisms. In: Marr JJ, Muller M, eds. Biochemistry and Molecular Biology of Parasites. London: Academic Press, 1995:147-160
72. Fridovich I. Superoxide anion radical, superoxide dismutases, and related matters. J Biol Chem 1997; 272:18515-18517.
73. Le Trant N, Meshnick SR, Kitchener K et al. Iron-containing superoxide dismutase from Crithidia fasciculata: Purification, characterisation, and similarity to leishmanial and trypanosomal enzymes. J Biol Chem 1983; 258:125-130.
74. Temperton NJ, Wilkinson SR, Meyer DJ et al. Overexpression of superoxide dismutase in Trypanosoma cruzi results in increased sensitivity to the trypanocidal agents gentian violet and benznidazole. Mol Biochem Parasitol 1998; 96:167-176.
75. Temperton NJ, Wilkinson SR, Kelly JM. Cloning of an Fe-superoxide dismutase gene homologue from Trypanosoma cruzi. Mol Biochem Parasitol 1996; 76:339-343.
76. Ismail SO, Paramchuk W, Skeiky YAW et al. Molecular cloning of two iron superoxide dismutase cDNAs from Trypanosoma cruzi. Mol Biochem Parasitol 1997; 86:187-197.
77. Parker MW, Blake CCF. Iron- and manganese- containing superoxide dismutases can be distinguished by analysis of their primary structures. FEBS Lett 1988; 229:377-382.
78. Yim MB, Chock PB, Stadtman ER. Copper, zinc superoxide dismutase catalyses hydroxyl radical production from hydrogen peroxide. Proc Natl Acad Sci USA 1990; 87:5006-5010.
79. Morello A. The biochemistry of the mode of action of drugs and the detoxification mechanisms in Trypanosoma cruzi. Comp Biochem Physiol 1988; 90C:1-12.
80. Amstad P, Moret R, Cerutti P. Glutathione peroxidase compenstates for the hypersensitivity of Cu,Zn-superoxide dismutase overproducers to oxidant stress. J Biol Chem 1994; 269:1606-1609.
81. Holmgren A. Thioredoxin and glutaredoxin systems. J Biol Chem 1989; 264:13963-13966.
82. Grant CM. Role of the glutathione/glutaredoxin and thioredoxin systems in yeast growth and response to stress conditions. Mol Microbiol 2001; 39:533-541.
83. Pedrajas JR, Kosmidou E, Miranda-Vizuete A et al. Identification and functional characterisation of a novel mitochondrial thioredoxin system in Saccharomyces cerevisiae. J Biol Chem 1999; 274:6366-6373.

84. Alphey MS, Leonard GA, Gourley DG et al. The high resolution crystal structure of recombinant Crithidia fasciculata tryparedoxin-1. J Biol Chem 1999; 274:25613-25622.
85. Reckenfelderbaumer N, Ludemann H, Schmidt H et al. Identification and functional characterization of thioredoxin from Trypanosoma brucei brucei. J Biol Chem 2000; 275:7547-7552.
86. Dormeyer M, Reckenfelderbaumer N, Ludemann H et al. Trypanothione-dependent synthesis of deoxyribonucleotides by Trypanosoma brucei ribonucleotide reductase. J Biol Chem 2001; 276:10602-10606.
87. Tetaud E, Fairlamb AH. Cloning, expression and reconstitution of the trypanothione-dependent peroxidase system of Crithidia fasciculata. Mol Biochem Parasitol 1998; 96:111-123.
88. Guerrero SA, Flohe L, Kalisz HM et al. Sequence, heterologous expression and functional characterization of tryparedoxin 1 from Crithidia fasciculata. Eur J Biochem 1999; 259:789-794.
89. Nogoceke E, Gommel DU, Kiess M et al. A unique cascade of oxidoreductases catalyses trypanothione-dependent peroxide metabolism in Crithidia fasciculata. Biol Chem 1997; 378:8220-8225.
90. Wilkinson SR, Meyer DJ, Kelly JM. Biochemical characterisation of a trypanosome enzyme with glutathione-dependent peroxidase activity. Biochem J 2000; 352:755-761.
91. Henderson GB, Fairlamb AH, Ulrich P et al. Trypanothione dependent peroxide metabolism in Crithidia fasciculata and Trypanosoma brucei. Mol Biochem Parasitol 1987; 24:39-45.
92. Penketh PG, Kennedy WP, Patton CL et al. Trypanosomatid hydrogen peroxide metabolism. FEBS Lett 1987; 221:427-431.
93. Chae HZ, Uhm TB, Rhee SG. Dimerization of thiol-specific antioxidant and the essential role of cysteine 47. Proc Natl Acad Sci USA 1994; 91:7022-7026.
94. Ellis HR, Poole LB. Roles for the two cysteine residues of AhpC in catalysis of peroxide reduction by alkyl hydroperoxide reductase from Salmonella typhimurium. Biochemistry 1997; 36:13347-13356.
95. Alphey MS, Bond CS, Tetaud E et al. The structure of reduced tryparedoxin peroxidase reveals a decamer and insight into reactivity of 2Cys-peroxiredoxins. J Mol Biol 2000; 300:903-916.
96. Wilkinson SR, Temperton NJ, Mondragon A et al. Distinct mitochondrial and cytosolic enzymes mediate trypanothione-dependent peroxide metabolism in Trypanosoma cruzi. J Biol Chem 2000; 275:8220-8225.
97. Lopez JA, Carvalho TU, de Souza W et al. Evidence for a trypanothione-dependent peroxidase system in Trypanosoma cruzi. Free Radical Biol Med 2000; 28:767-772.
98. Smith K, Opperdoes FR, Fairlamb AH. Subcellular distribution of trypanothione reductase in bloodstream and procyclic forms of Trypanosoma brucei. Mol Biochem Parasitol 1991; 48:109-112.
99. Martensson J, Lai JC, Meister A. High-affinity transport of glutathione is part of a multicomponent system essential for mitochondrial function. Proc Natl Acad Sci USA 1990; 87:71185-7189.
100. Chen Z, Lash LH. Evidence for mitochondrial uptake of glutathione by dicarboxylate and 2-oxoglutarate carriers. J Pharmacol Exp Ther 1998; 285:608-618.
101. Watabe S, Hiroi T, Yamamoto Y et al. SP-22 is a thioredoxin-dependent peroxide reductase in mitochondria. Eur J Biochem 1997; 249:52-60.
102. Spyrou G, Enmark E, Miranda-Vizuete A et al. Cloning and expression of a novel mammalian thioredoxin. J Biol Chem 1997; 272:2936-2941.
103. Ursini F, Maiorino M, Aumann KD et al. Diversity of glutathione peroxidases. Methods Enzymol 1995; 252:38-53.
104. Rocher C, Lalanne JL, Chaudiere J. Purification and properties of a recombinant sulfur analog of murine selenium-glutathione peroxidase. Eur J Biochem 1992; 205:955-960.
105. Maiorino M, Aumann KD, Brigelius-Flohe R et al. Probing the presumed catalytic triad of selenium-containing peroxidases by mutational analysis of phospholipid hydroperoxide glutathione peroxidase (PHGPx). Biol Chem Hoppe-Seyler 1995; 376:651-660.
106. Bjornstedt M, Xue J, Huang W et al. The thioredoxin and glutaredoxin systems are efficient electron donors to human plasma glutathione peroxidase. J Biol Chem 1994; 269:29382-29384.

Chapter 5

Ca^{2+} Signaling in the Invasion of Mammalian Cells by *Trypanosoma cruzi*

Silvia N.J. Moreno and Roberto Docampo

Summary

In order to replicate in the mammalian host, *Trypanosoma cruzi* must invade host cells. Changes in the intracellular Ca^{2+} concentration ($[Ca^{2+}]_i$) of *T. cruzi* and tissue culture cells during their interaction have been demonstrated. When formation of Ca^{2+} transients is prevented by intracellular Ca^{2+} chelators, in either the parasite or the host cells, a decrease in host invasion is observed. This reveals the importance of $[Ca^{2+}]_i$ in the process of parasite-host cell interaction. Different stimuli for the occurrence of these $[Ca^{2+}]_i$ changes, such as attachment of the parasites to the host cells, or membrane proteins and soluble factors of parasite or host cell origin were shown to be responsible for these changes. Ca^{2+} influx or Ca^{2+} release from intracellular stores have both been suggested as the sources for these Ca^{2+} changes. Several *T. cruzi* stages, strains, and host cells have been used in these studies and there is therefore evidence of multiple mechanisms of Ca^{2+} signaling in the cells involved.

Introduction

Intracellular Ca^{2+} $[Ca^{2+}]_i$ is a ubiquitous signal responsible for controlling numerous cellular processes and *Trypanosoma cruzi* has developed ways to manipulate Ca^{2+} signaling to induce its own internalization by host cells.

T. cruzi is able to invade excitable and nonexcitable cells as well as phagocytic and nonphagocytic cells. In addition, three different parasite stages are able to invade mammalian cells: (1) the bloodstream trypomastigote, that can also be obtained after in vitro cultivation of parasites in tissue culture cells; (2) the metacyclic trypomastigotes present in the insect vector and that can also be obtained in vitro after differentiation from epimastigotes in axenic medium; and (3) the amastigote, that is the intracellular form. Amastigotes can be released to the extracellular medium of tissue culture cells or to the bloodstream, before their differentiation to trypomastigotes, because of premature rupture of the host cell. Because of the low parasitemias in animals and the limited yield of metacyclics from the insect vector, most studies on Ca^{2+} signaling have been done with either tissue culture-derived trypomastigotes or metacyclic trypomastigotes obtained in vitro. No studies have been reported using *T. cruzi* amastigotes.

The use of in vitro derived parasites has two potential disadvantages. One is that these stages might not behave as the parasites obtained from animals, and second, that these stages are obtained after cultivation in very rich culture media and some growth factors or other

Molecular Mechanisms in the Pathogenesis of Chagas Disease, edited by John M. Kelly.

materials from the culture medium not present in the blood or in the intestine of the vector, can remain attached to the parasites or to the tissue culture cells, as recent reports have shown.[1] On the other hand, it is also possible that other factors that are normally attached to the bloodstream forms or to the metacyclic trypomastigotes in vivo are not present in those in vitro stages. A further complication is the recently described existence of two lineages of *T. cruzi* that have evolved independently for a long time.[2] The epidemiological, biological and biochemical attributes of these groups are sufficiently distinct that it has been proposed to confer the status of taxonomic species to each group.[2] *T. cruzi* group I (such as the Tulahuén, Silvio X10, Colombiana or Dm28 strains) has a predominantly sylvatic cycle while *T. cruzi* group II (such as Y and CL strains) has a predominantly domestic cycle. It is not yet known to what extent there are differences in the pathogenic potential of the members of the two groups or whether they give rise to different manifestations of Chagas disease, although differences appear to exist in the expression of some surface proteins.[2]

Since the mechanisms and roles of Ca^{2+} signaling in these different cells (host and parasites) could be different, it is not surprising that apparently conflicting results have been found when tissue culture cell lines of different origins, or different stages of the parasite, have been used to investigate the role of Ca^{2+} signaling in invasion by *T. cruzi*.

Ca^{2+} signaling involves the mobilization of Ca^{2+} from two sources: intracellular stores and the extracellular medium. In most instances the release of intracellular Ca^{2+} and the activation of Ca^{2+} entry across the plasma membrane are coordinated processes.[3] In excitable mammalian cells, such as in muscle cells, the primary signal (membrane depolarization) activates Ca^{2+} entry across the plasma membrane, and this Ca^{2+} signal is amplified and propagated by a mechanism of Ca^{2+}-induced Ca^{2+} release (CICR) from the sarcoplasmic reticulum through the activation of ryanodine receptors.[4] In nonexcitable mammalian cells, such as endothelial cells, the primary signal (usually a hormone or growth factor) could lead to the stimulation of a phospholipase C that hydrolyzes phosphatidylinositol 4,5-disphosphate (PIP_2) to generate inositol 1,4,5-trisphosphate (IP_3) and diacylglycerol. IP_3 diffuses to the intracellular stores (endoplasmic reticulum) and causes the release of Ca^{2+} into the cytoplasm through the activation of IP_3 receptors.[5] This release of intracellular Ca^{2+} can in some manner signal the activation of Ca^{2+} entry, a process known as capacitative Ca^{2+} entry.[3,6] Some mammalian cells, such as smooth muscle and neuroendocrine cells, can utilize both of these pathways. In addition, a number of newly discovered second messengers, such as for example cyclic adenosine diphosphate ribose (cADPr)[7], sphingosine-1-phosphate,[8] and nicotinic acid adenine dinucleotide phosphate (NAADP)[9] have been described as being able to release or modulate the release of intracellular Ca^{2+} from different cells.

In *T. cruzi*, as in mammalian cells, $[Ca^{2+}]_i$ increases could involve the mobilization of Ca^{2+} from intracellular stores or the extracellular medium. However, little is known about signals and receptors involved in these mechanisms. The most important intracellular calcium storage compartment in different stages of *T. cruzi* is the acidocalcisome, an acidic vacuole rich in polyphosphates, calcium, sodium, magnesium and zinc.[10-15] There is also evidence for the involvement of the mitochondria,[16-18] the plasma membrane,[19,20] and the endoplasmic reticulum[21,22] in Ca^{2+} homeostasis. However, although an inositol-1,4,5-trisphosphate/diacylglycerol signaling pathway is present in different stages of the parasite[23-25] it is not known how Ca^{2+} is released from either the acidocalcisome or the endoplasmic reticulum.[15] There is also no firm evidence of the operation of a "capacitative" calcium uptake mechanism. The mechanism of Ca^{2+} entry from the extracellular medium has not been investigated except for a recent report in amastigotes suggesting its stimulation by fatty acids.[26]

Studies with Tissue Culture-Derived Trypomastigotes

Intracellular Calcium Increase in Tissue Culture Cells

Early work[27,28] described increases in the $[Ca^{2+}]_i$ of host cells during their interaction with *T. cruzi.* However, these changes were measured after prolonged times of interaction with the parasites and were probably not involved with the invasion. Changes in $[Ca^{2+}]_i$ of host cells were measured by loading the cells with acetoxymethyl ester derivatives of fluorescent Ca^{2+} indicators such as fura 2 or fluo 3. Once inside the cells these dyes are cleaved by nonspecific esterases and are able to respond to $[Ca^{2+}]_i$ changes. Morris et al[27] found an increase in $[Ca^{2+}]_i$ of fura 2-loaded umbilical vein endothelial cells (from 55 ± 18 nM to 110 ± 20 nM) four days after infection with trypomastigotes (Tulahuén strain), well after transformation of trypomastigotes into amastigotes, when amastigotes were actively duplicating in the host cells. Low et al[28] found an increase in $[Ca^{2+}]_i$ in fura 2-loaded BSC-1 cells (African monkey *Cercopitherus aethiops* kidney cell line) one day (the earliest time point examined) after infection with trypomastigotes (Brazil strain) followed by a subsequent decrease up to about five days after infection to a minimum of about 7 nM.

Osuna et al[29] were the first to find early changes in host $[Ca^{2+}]_i$ upon interaction with *T. cruzi* (Table 1). They found a three-fold increase in $[Ca^{2+}]_i$ of aequorin-loaded HeLa cells after one-hour interaction with trypomastigotes (Venezuela strain). Treatment of the host cells with EDTA, used as a Ca^{2+} chelator, before and during their interaction with trypomastigotes reduced the parasitization percentages, thus suggesting a role for Ca^{2+} in invasion.[29] More recent reports[30,31] indicated that the parasite was able to trigger even earlier Ca^{2+} signals in the host cells. Tardieux et al,[30] using fluo 3-loaded normal rat kidney (NRK) fibroblasts, found that trypomastigotes (Y strain) induced an increase in $[Ca^{2+}]_i$ in the host cells within 200 sec of interaction. *T. cruzi* trypomastigotes or their isolated membranes but not epimastigotes, induced repetitive cytosolic-free Ca^{2+} transients in individual cells. Depletion of host cell Ca^{2+} before addition of the parasites, by exposure of the fibroblasts to the Ca^{2+} ionophore A23187 or the intracellular chelator MAPTA-AM in Ca^{2+}-free medium, further inhibited trypanosome entry suggesting a role for Ca^{2+} signaling in the host during *T. cruzi* invasion.[30] When using L_6E_9 myoblasts (a murine myoblast cell line derived from line L_6) loaded with fura 2, an early increase in host $[Ca^{2+}]_i$ occurred upon contact with trypomastigotes (Y strain).[31] In addition, host $[Ca^{2+}]_i$ was elevated around intracellular parasites.[31] The apparent spreading of the elevated $[Ca^{2+}]_i$ from the region surrounding the nonmotile parasites into the remainder of the cell, suggested that it occurred after internalization.[31] Pretreatment of rat heart myoblasts with Ca^{2+} chelators (BAPTA, Quin 2) also inhibited trypomastigote (Tulahuén strain) invasion, while treatment with the Ca^{2+} ionophore ionomycin enhanced invasion.[32]

Stimulus for $[Ca^{2+}]_i$ Increase in the Host Cells

Membrane and soluble factors either from the parasite or from the host (attached to the parasites or to the host cells) have been involved in the generation of $[Ca^{2+}]_i$ increases in host cells during their interaction with *T. cruzi* (Table 2).

Tardieux et al[30] first postulated that a trypomastigote membrane factor triggered cytosolic-free Ca^{2+} transients in NRK (normal rat kidney fibroblasts) cells. This was because these Ca^{2+} transients were elicited by intact or heat-killed trypomastigotes, or by membrane-fractions isolated from them. The proteic nature of the stimulus was suggested by experiments in which treatment of the trypomastigote membranes with trypsin abolished their capacity to trigger Ca^{2+} signals in NRK cells.[30] Contact of trypomastigotes with L_6E_9 myoblasts[31,33] (see below) or NRK cells[34] was also postulated to be involved in rises in $[Ca^{2+}]_i$ in the host cells.

***Table 1. Early increases in intracellular Ca^{2+} concentration in mammalian cells upon interaction with live* T. cruzi**

T. cruzi Strains	Tissue Culture Cell	Reference
Tissue culture-derived trypomastigotes (TCT)		
Venezuela	HeLa	29
Y	NRK	30, 34
	L_6E_9	31
Dm28c	HUVEC	37
	CHO-B_2R	37
Metacyclic trypomastigotes (MT, culture forms)		
G, CL	HeLa	40, 42
RA	Peritoneal macrophages	41

NRK are normal rat kidney fibroblasts; L_6E_9 are myoblasts; HUVEC are human umbilical cord endothelial cells; CHO-B_2R are Chinese hamster ovary cells overexpressing the bradykinin 2 receptor.

It was also shown that trypomastigotes express a soluble Ca^{2+} signaling activity (trypanosome soluble factor, TSF) that is generated through the action of a cytosolic oligopeptidase B.[33-36] A soluble extract of trypomastigotes (Y strain) was able to induce elevations in $[Ca^{2+}]_i$ in NRK fibroblasts, Chinese hamster lung (Dede) cells, Madin-Darby canine kidney cells (MDCK), A7 cells derived from a human malignant melanoma cell line, Chinese hamster ovary (CHO) cells and African green monkey (CV-1) cells.[35] $[Ca^{2+}]_i$ increase in NRK cells was prevented by preincubation of the parasite extracts with some protease inhibitors (leupeptin, chymostatin, Z-FR-FMK, Ac-RR-CMK, or antipain) but not by others (aprotinin, E-64, phenylmethysulfonyl fluoride, pepstatin, SBTI, cystatin). A 120 kDa alkaline peptidase was found in the trypomastigote extracts and postulated to cleave a trypomastigote-specific factor that became activated and able to induce Ca^{2+}-signaling in the host cells.[35] Although this factor was not identified, it was found to be sensitive to *N*-ethylmaleimide and was able to bind to heparin.[35] A similar increase in $[Ca^{2+}]_i$ was induced by the same soluble factor in primary cardiac myocytes.[36] The gene encoding the oligopeptidase was cloned, sequenced and expressed in bacteria.[33] Antibodies against the recombinant enzyme were shown to inhibit peptidase activity and Ca^{2+} signaling induced in NRK cells by trypomastigote extracts.[33] Deletion of the gene encoding oligopeptidase B resulted in decreased cell invasion and establishment of infection in mice.[33] It was found that trypomastigotes with a single and double knock out of the oligopeptidase gene were still able to invade L_6E_9 myoblasts and HeLa cells. This residual invasion capacity was prevented by preincubation of L_6E_9 myoblasts with MAPTA-AM to buffer intracellular free Ca^{2+} suggesting that the residual host cell invasion by oligopeptidase B mutants was also dependent on Ca^{2+} transients.[33] The soluble Ca^{2+} signaling activity of oligopeptidase B null mutants was reconstituted with recombinant oligopeptidase B. However, soluble extracts from double knock out mutants retained the ability to trigger Ca^{2+} transients in NRK fibroblasts and this activity was abolished by several protease inhibitors that inhibited oligopeptidase B (leupeptin, TLCK, Z-FR-FMK) and by an additional inhibitor (Z-FA-FMK) that did not inhibit the oligopeptidase B.[33] This led the authors to speculate that another peptidase, sensitive to Z-FA-FMK, was present in trypomastigote extracts and was also able to generate a Ca^{2+} agonist for the NRK cells. However, since Z-FA-FMK failed to reduce the residual invasion of L_6E_9 myoblasts by oligopeptidase B null mutants, the authors suggested that the generation of the

***Table 2. Stimulus for Ca^{2+} increase in the host cells upon interaction with* T. cruzi**

***T. cruzi* Strain**	**Tissue Culture Cell**	**Reference**
Cell contact		
Venezuela (TCT)	HeLa	29
Y	L_6E_9	31
	NRK	34
G, CL (MT)	HeLa	40
RA (MT)	Peritoneal macrophages	41
Parasite soluble factor		
Y (TCT)	NRK	33,34,35
	Dede	35
	MDCK	35
	CHO	35
	A7	35
	CV-1	35
	Cardiac myocytes	36
	Xenopus laevis oocytes	39
Parasite protein		
Y (TCT)	NRK	30
G, CL (MT)	HeLa	40,42
Host soluble factor		
Dm28c (TCT)	HUVEC	37
	CHO-B_2R	37

Dede are Chinese hamster lung cells; MDCK are Mardin-Darby canine kidney cells; CHO are Chinese hamster ovary cells; A7 are cells derived from a human malignant melanoma cell line; CV-1 are African green monkey cells, *Xenopus laevis* oocytes microinjected with mRNA from NRK fibroblasts. MT, metacyclic trypomastigotes; TCT, tissue culture-derived trypomastigotes.

soluble Ca^{2+} agonist by the additional protease/peptidase activity in soluble parasite extracts is not physiologically relevant to the mechanism of *T. cruzi* invasion.[33] Attachment of the parasites to the host cells or release of an additional previously undetected factor, were proposed as the more likely sources of Ca^{2+} and to be responsible for the residual invasion of oligopeptidase B null mutants.[33] Evidence in favor of a localized elevation of $[Ca^{2+}]_i$ at the site of trypomastigote (Y strain) attachment, and that these signals were less efficiently generated by the oligopeptidase B null mutants was also presented.[34]

A novel mechanism of Ca^{2+} signaling by *T. cruzi* (Dm28c clone) was recently proposed.[37] Evidence was provided that the main cysteine protease of *T. cruzi*, cruzipain, is released to the extracellular medium where it is able to cleave kininogen that remains associated with the cell surfaces of target cells and/or is displayed by the parasites. This leads to the formation of kinin-signaling peptides that increase Ca^{2+} and favor invasion of the host cells (human primary umbilical vein endothelial cells (HUVEC) or Chinese hamster ovary cells overexpressing the B_2 type of bradykinin receptors).[37]

Table 3. Sources of Ca^{2+} increase induced by interaction with* T. cruzi *trypomastigotes or parasite-derived material

T. cruzi Strain	Tissue Culture Cell	Reference
Ca^{2+} influx		
Venezuela (TCT)	HeLa	29
G, CL (MT)	HeLa	40
RA (MT)	Peritoneal macrophages	41
Intracellular stores		
Y (TCT)	NRK	30, 34, 35, 38, 39
	Cardiac myocytes	36
G, CL (MT)	HeLa	40

Cell lines are as in Tables 1 and 2.

Sources of Ca^{2+} in the Host Cells

Both Ca^{2+} influx and Ca^{2+} release from intracellular stores have been implicated as the source of $[Ca^{2+}]_i$ increase in host cells (Table 3).

Osuna et al[29] found that treatment of the HeLa cells with verapamil (20 μM for 1 hour), a calcium channel blocker, before and during their interaction with trypomastigotes reduced the parasitization percentages by *T. cruzi* suggesting a role for Ca^{2+} influx in parasite invasion. Since *T. cruzi* invasion was inhibited by pretreatment of the host cells with the channel blockers $NiCl_2$ (5 mM for 15 min) and verapamil (100 μM for 30 min) and NRK cells did not respond with Ca^{2+} signals when exposed to $NiCl_2$, a requirement for Ca^{2+} influx was considered a necessary component of the parasite-induced signaling process.[30]

Evidence was also provided for a G-protein-coupled pathway for Ca^{2+} release from intracellular stores.[30,36,38,39] It was postulated that the TSF, acting like a hormone, would stimulate a G-protein-coupled receptor that signals via phosphoinositide hydrolysis leading to Ca^{2+} mobilization from intracellular stores. In favor of this hypothesis the authors found that: (1) $[Ca^{2+}]_i$ increase in NRK cells[30] or primary cardiac myocytes[36] was inhibited by preincubation of the cells with pertussis toxin (0.4 μg/ml for 4 hours, NRK cells, or 6 hours, cardiac myocytes); (2) trypomastigote soluble extracts induced a small increase in inositol-1,4,5-trisphosphate formation in NRK cells;[38] (3) incubation of NRK cells[30] or primary cardiac myocytes[36] with thapsigargin (0.5 μM for 30 min), an inhibitor of sarcoplasmic/endoplasmic reticulum (SERCA)-ATPases, prevented Ca^{2+} increases and invasion suggesting that Ca^{2+} was released from intracellular stores; (4) treatment of cardiac myocytes with ryanodine (2.0 μM for 15 min), a ryanodine receptor inhibitor, also prevented these Ca^{2+} increases and cell invasion, while pretreatment with Ca^{2+} channel blockers (cadmium, 200 μM; nisoldipine, 10 μM; verapamil, 100 μM) for 15 min or in the absence of extracellular Ca^{2+} did not inhibit this response;[36] (5) a response of identical characteristics to that found in NRK cells (inhibited by pertussis toxin, thapsigargin, and oligopeptidase B inhibitors) was found in *Xenopus laevis* oocytes injected with mRNA from normal NRK cells.[39]

Since oligopeptidase B null mutants were still invasive and exhibited a Ca^{2+} response that was not affected by pretreatment of L_6E_9 myoblast with thapsigargin, a second oligopeptidase B-independent pathway was postulated to be stimulated by trypomastigotes.[33] Ca^{2+} influx from the extracellular medium, either by mechanical stimulation after attachment of the parasites to the host cells or by release of an additional previously undetected factor, was proposed as the more likely source of Ca^{2+} responsible for the residual invasion of oligopeptidase B null mutants.[33]

Studies with Metacyclic Trypomastigotes

Studies with this parasite stage have been less frequent. Contact of live metacyclic trypomastigotes with HeLa cells[40] or peritoneal macrophages[41] preloaded with fura 2 was able to raise host $[Ca^{2+}]_i$ (Table 1). Depletion of host intracellular Ca^{2+} (and probably also parasite Ca^{2+}) by a pretreatment of the HeLa cells for 15 min with a combination of ionophore A23187 and absence of extracellular calcium (no calcium added to the medium), and then continuation of this treatment during their interaction with trypomastigotes (3 hours), drastically reduced invasion of HeLa cells.[40] Ca^{2+} increase and invasion was also prevented by preincubation of macrophages with the calcium chelator BAPTA-AM (5 μM for 2 hours).[40]

The source of the $[Ca^{2+}]_i$ increase in HeLa cells incubated with metacyclic trypomastigotes (G and CL strains) was postulated to be both the extracellular medium and the intracellular stores (Table 3). This was because: (1) parasite invasion was reduced when their interaction with HeLa cells was carried out in the presence of $CdCl_2$ (1 μM), which was used as a nonspecific calcium-channel antagonist; and (2) incubation of the cells with thapsigargin (1 μM for 3 hours) reduced trypomastigote entry into the host cells.[40] Macrophage invasion by metacyclic trypomastigotes was also prevented by preincubation of the cells with the calcium channel antagonist methoxyverapamil (10 μM for 1 hour).[41]

Ca^{2+} signaling in HeLa cells was also detected with sonicated parasite preparations but, in contrast to studies using tissue culture-derived trypomastigotes,[33-36] $[Ca^{2+}]_i$ increases were observed in the presence or absence of protease inhibitors.[40] Low, but not high, concentrations of purified preparations of two metacyclic surface glycoproteins, gp35/50 and gp82 also stimulated $[Ca^{2+}]_i$ increase in HeLa cells.[40] The same group later found that the Ca^{2+} response triggered in HeLa cells by gp82 was significantly higher than that induced by gp35/50 or gp90.[42] Interestingly, gp90 was found to be present mainly in strains of the *T. cruzi* group I but not in strains of the *T. cruzi* group II.[42] Parasite lysates or metacyclic trypomastigote membranes were also able to elicit a $[Ca^{2+}]_i$ increase in macrophages but no protease inhibitors were tested in this study.[41] On the other hand, oligopeptidase B single and double knockout metacyclic trypomastigote mutants (Y strain) were less effective than wild type metacyclics in invading NRK fibroblasts and L_6E_9 myoblasts, supporting a role for an oligopeptidase B pathway in invasion of host cells by this stage.[33]

Calcium Signaling in Trypomastigotes during Host Cell Invasion

In addition to an increase in intracellular Ca^{2+} in the host cells, an increase in cytosolic Ca^{2+} was also demonstrated to occur in tissue culture-derived trypomastigotes (Y strain) after their association with the host cells.[31] Buffering cytosolic Ca^{2+} of trypomastigotes by intracellular Ca^{2+} chelators (BAPTA or Quin 2) resulted in an inhibition of cellular invasion.[31] Yakubu et al[32] also observed that pretreatment of both, tissue-culture derived and bloodstream trypomastigotes (Tulahuén strain), with the same intracellular Ca^{2+} chelators decreased their infectivity while treatment with the Ca^{2+} ionophore ionomycin, which elevated $[Ca^{2+}]_i$ in trypomastigotes, significantly enhanced the infective capacity of the parasites. The mechanism and sources of the increased $[Ca^{2+}]_i$ in trypomastigotes are unknown. A report has shown that host cell components (sonicated extracts of HeLa cells) and monoclonal antibodies against *T. cruzi* surface molecules (gp35/50 and gp82) induced small Ca^{2+} responses in fura 2-loaded metacyclic trypomastigotes (G and CL strain). The authors postulated that the antibodies could be mimicking the host cell receptors for *T. cruzi* proteins.[42]

Why Are Increases in $[Ca^{2+}]_i$, in Both the Host Cell and the Parasite, Needed for Cell Invasion?

It has been postulated that trypomastigotes enter nonphagocytic cells by a unique mechanism, distinct from phagocytosis. Invasion appears to be an active process facilitated by disruption of host cell actin microfilaments, and involving recruitment and fusion of host lysosomes at the site of parasite entry.[43,44] Cytosolic-free Ca^{2+} transients have been postulated to be required for local rearrangement of the cortical actin cytoskeleton allowing lysosome access to the plasma membrane, and lysosome fusion at the site of trypanosome entry.[45] The following evidence, obtained during interaction of tissue culture-derived trypomastigotes (Y strain) with different tissue culture cells, is in favor of this hypothesis: (1) lysosomes from NRK cells were observed to aggregate at the sites of trypomastigote attachment and to fuse with the parasitophorous vacuole at early stages of its formation.[43] Direct migration of lysosomes of L_6E_9 myoblasts was observed by time-lapse video-enhanced microscopy only when their original position was at less than 11-12 μm from the parasite entry site;[44] (2) experimentally induced microtubule-dependent movement of lysosomes from the perinuclear area to the cell periphery (by pretreatment of NRK cells with: 10 μM brefeldin A for 30 min; 1 mM dibutyryl cyclic AMP for 30 min; sodium acetate, pH 6.6, for 15 min), or treatment with different concentrations of the actin microfilament disruptor cytochalasin D (50 nM to 10 μM) for different times (1 to 30 min) enhanced entry, while conditions that depleted NRK cells of peripheral lysosomes (alkalinization with 40 mM NH_4Cl), or interfered with lysosomal fusion capacity (sucrose loading of lysosomes, mitosis) inhibited invasion;[43] (3) depletion of peripheral lysosomes by microinjection of NRK cells with antibodies against the cytoplasmic domain of lgp 120, a lysosomal protein, treatment of L_6E_9 myoblasts or NRK cells with the microtubule-binding drugs nocodazole, colchicine, vinblastine and taxol, or microinjection of NRK cells with antibodies to the heavy chain of kinesin, which blocked the acidification-induced, microtubule-dependent redistribution of lysosomes to the host cell periphery, also reduced trypomastigote entry;[44] (4) lysosomes behave as Ca^{2+}-regulated exocytic vesicles in several cell lines of fibroblasts and epithelial cells;[46] (5) stimulation of NRK cells with isoproterenol, a β-adrenergic agonist that activates adenylyl cyclase, enhanced Ca^{2+}-dependent lysosome exocytosis and cell invasion by trypomastigotes, which were also able to trigger a small elevation in host cell cAMP levels;[47] (6) expression of the lysosomal membrane glycoprotein-1 (Lamp-1) at the cell surface rendered CHO cells more susceptible to trypomastigote invasion in a microtubule-dependent fashion and Ca^{2+}-triggered exocytosis of lysosomes was enhanced in these cells.[48]

Although the evidence described above is very compelling several observations suggest that this postulated mechanism of trypomastigote entry into nonphagocytic cells might not be universal. For example, Schenkman et al[49] reported that treatment of MDCK cells with cytochalasin D (an actin filament disruptor) (10 μM for 30 min) inhibited tissue culture-derived trypomastigote entry, while the same treatment did not significantly alter penetration into HeLa cells. HeLa cells were shown to extend and internalize pseudopodia around actively invading metacyclic trypomastigotes.[50] These membrane protrusions were shown to be actin-rich structures[51] and this is in contrast to the invasion of NRK cells which was shown to be independent of the host cell microfilament system.[43] The cytoskeleton of heart muscle cells (HMC) also participates in their interaction with metacyclic trypomastigotes, [52] and in contrast to results obtained with tissue culture-derived trypomastigotes and NRK cells,[43] incubation of HMC cells with cytochalasin D (5 μg/ml for 30 or 60 min) inhibited host cell invasion.[51] Furthermore, tissue culture derived trypomastigotes (Y strain) did not induce significant changes in the distribution of actin filaments, microtubules or lysosomes in Vero cells (kidney fibroblasts of African green monkey) during the first 48 hours of infection.[53]

Since cytochalasin D was shown to inhibit invasion of wild type and oligopeptidase B null mutant trypomastigotes into J774 macrophages, it was suggested that the residual invasion mechanism of the signaling-deficient oligopeptidase B null trypomastigotes may also involve lysosome recruitment, which is facilitated by disruption of the actin cytoskeleton.[34] No studies have been reported concerning the need for $[Ca^{2+}]_i$ increase in trypomastigotes for parasite invasion although a possible role in the processing of the trypanosome soluble factor has been suggested.[54]

Concluding Remarks

In conclusion, early changes in $[Ca^{2+}]_i$ of different tissue culture cells have been found upon interaction with tissue culture-derived or metacyclic trypomastigotes of *T. cruzi*. Early changes in the $[Ca^{2+}]_i$ of tissue culture-derived trypomastigotes have also been detected during this interaction. It has been shown that the changes in both the parasites and the host cells are important signaling mechanisms for invasion. Both, the extracellular medium and the intracellular stores of tissue culture cells have been identified as the source of Ca^{2+}, and attachment of the parasites, membrane proteins and soluble factors of parasite or host cell origin have all been implicated in these signaling processes. Cytoskeletal modifications appear to be stimulated by the Ca^{2+} changes in some, but apparently not all of the infected cells. Differences in the tissue culture cell, parasite stage, parasite strain, and the culture medium are probably responsible for some of the different results described.

Acknowledgements

Work in the Laboratory of Molecular Parasitology was supported by a grant from the National Institutes of Health (AI-23259).

References

1. Scharfstein J, Schmitz V, Morandi V et al. Host cell invasion by Trypanosoma cruzi is potentiated by activation of bradikynin B_2 receptors. J Exp Med 2000; 192:12889-1299.
2. Zingales B. Revisiting Trypanosoma cruzi: the meaning of the two major groups. Two species of the same parasite? Mem Instituto Oswaldo Cruz 2000; 95(Suppl. II):10-12.
3. Putney JW, Bird GStJ. Calcium mobilization by inositol phosphates and other intracellular messengers. Trends Endocrinol Methods 1994; 5:256-260.
4. Fabiato A. Calcium-induced release of calcium from the cardiac sarcoplasmic reticulum. Am J Physiol 1983; 245:C1-C4.
5. Berridge, MJ. Inositol trisphophate and diacylglycerol: two interacting second messengers. Annu Rev Biochem 1987; 56:159-193.
6. Putney Jr JW. Capacitative calcium entry revisited. Cell Calcium 1990; 11:611-624.
7. Lee HC. Cyclic ADP-ribose: a new member of a super family of signalling cyclic nucleotides. Cellular Signalling 1994; 6:591-600.
8. Mattie M, Brooker G, Spiegel S. Sphingosine-1-phosphate, a putative second messenger mobilizes calcium from internal stores via an inositol trisphosphate-independent pathway. J Biol Chem 1994; 269:3181-3188.
9. Lee HC, Aarhus R. A derivative of NADP mobilizes calcium stores insensitive to inositol trisphosphate and cyclic ADP-ribose. J Biol Chem 1995; 270:2152-2157.
10. Docampo R, Scott DA, Vercesi AE et al. Intracellular Ca^{2+} storage in acidocalcisomes of Trypanosoma cruzi. Biochem J 1995; 310:1005-1012.
11. Scott DA, Docampo R, Dvorak JA et al. In situ compositional analysis of acidocalcisomes in Trypanosoma cruzi. J Biol Chem 1997; 272:28020-28029.
12. Lu H-G, Zhong L, de Souza W et al. Ca^{2+} content and expression of an acidocalcisomal calcium pump are elevated in intracellular forms of Trypanosoma cruzi. Mol Cell Biol 1998; 18:2309-2323.
13. Scott DA, Docampo R. Characterization of isolated acidocalcisomes of Trypanosoma cruzi. J Biol Chem 2000; 275:24215-24221.

14. Docampo, R, Moreno, SNJ. Acidocalcisome: a novel Ca^{2+} storage compartment in trypanosomatids and apicomplexan parasites. Parasitol Today 1999; 15:443-448.
15. Docampo R, Moreno SNJ. The acidocalcisome. Mol Biochem Parasitol 2001; 33:151-159.
16. Docampo R, Vercesi AE. Ca^{2+} transport by coupled Trypanosoma cruzi mitochondria in situ. J Biol Chem 1989; 264:108-111.
17. Docampo R, Vercesi AE. Characteristics of Ca^{2+} transport by Trypanosoma cruzi mitochondria in situ. Arch Biochem Biophys 1989; 272:122-129.
18. Vercesi AE, Bernardes CF, Hoffmann ME et al. Digitonin permeabilization does not affect mitochondrial function and allows the determination of the mitochondrial membrane potential of Trypanosoma cruzi in situ. J Biol Chem 1991; 266:14431-14434.
19. Benaim G, Losada S, Gadelha FR et al. A calmodulin-activated (Ca^{2+}-Mg^{2+})-ATPase is involved in Ca^{2+} transport by plasma membrane vesicles from Trypanosoma cruzi. Biochem J 1991; 280:715-720.
20. Benaim G, Moreno SNJ, Hutchinson G et al. Characterization of the plasma membrane calcium pump from Trypanosoma cruzi. Biochem J 1995; 306:299-303.
21. Vercesi AE, Hoffmann ME, Bernardes CF et al. Regulation of intracellular calcium homeostasis in Trypanosoma cruzi. Effects of calmidazolium and trifluoperazine. Cell Calcium 1991; 12:361-369.
22. Docampo R. Calcium homeostasis in Trypanosoma cruzi. Bio. Res 1993; 26:189-196.
23. Docampo R, Pignataro OP. The inositol phosphate/diacylglycerol signalling pathway in Trypanosoma cruzi. Biochem J 1991; 275:407-411.
24. Moreno SNJ, Vercesi AE, Pignataro OP et al. Calcium homeostasis in Trypanosoma cruzi amastigotes. Presence of inositol phosphates and lack of an inositol 1,4,5-trisphosphate-sensitive calcium pool. Mol Biochem Parasitol 1992; 52:251-262.
25. Docampo R, Moreno SNJ, Vercesi AE. Effect of thapsigargin on calcium homeostasis in Trypanosoma cruzi trypomastigotes and epimastigotes. Mol Biochem Parasitol 1993; 59:305-314.
26. Catisti R, Uyemura SA, Docampo R et al. Calcium mobilization by arachidonic acid in trypanosomatids. Mol Biochem Parasitol 2000; 105:261-271.
27. Morris SA, Tanowitz H, Hatcher V et al. Alterations in intracellular calcium following infection of human endothelial cell with Trypanosoma cruzi. Mol Biochem Parasitol 1988; 29:213-221.
28. Low HP, Paulin JJ, Keith CH. Trypanosoma cruzi infection of BSC-1 fibroblast cells causes cytoskeletal disruption and changes in intracellular calcium levels. J Protozool 1992; 39:463-470.
29. Osuna A, Castanys S, Rodriguez-Cabezas MN et al. Trypanosoma cruzi: calcium ion movement during internalization in host cell HeLa cells. Int J Parasitol 1990; 20:673-676.
30. Tardieux I, Nathanson MH, Andrews NW. Role in host cell invasion of Trypanosoma cruzi-induced cytosolic free Ca^{2+} transients. J Exp Med 1994; 179:1017-1022.
31. Moreno SNJ, Silva J, Vercesi AE et al. Cytosolic free calcium elevation in Trypanosoma cruzi is required for cell invasion. J Exp Med 1994; 180:1535-1540.
32. Schettino MS, Majumder S, Kierszenbaum F. Regulatory effect of the level of free Ca^{2+} of the host cell on the capacity of Trypanosoma cruzi to invade and multiply intracellularly. J Parasitol 1995; 81:597-602.
33. Caler EV, Vaena de Avalos S, Haynes PA et al. Oligopeptidase B-dependent signaling mediates host cell invasion by Trypanosoma cruzi. EMBO J 1998; 17:4975-4986.
34. Caler EV, Morty RE, Burleigh BA et al. Dual role of signaling pathways leading to Ca^{2+} and cyclic AMP elevation in host cell invasion by Trypanosoma cruzi. Infect Immun 2000; 68:6602-6610.
35. Burleigh BA, Andrews NW. A 120-kDa alkaline peptidase from Trypanosoma cruzi is involved in the generation of a novel Ca^{2+}-signaling factor for mammalian cells. J Biol Chem 1995; 270:5172-5180.
36. Barr SC, Han W, Andrews NW et al. A factor from Trypanosoma cruzi induces repetitive cytosolic free Ca^{2+} transients in isolated primary canine cardiac myocytes. Infect Immun 1996; 64:1770-1777.
37. Scharfstein J, Schmitz V, Morandi V et al. Host cell invasion by Trypanosoma cruzi is potentiated by activation of bradykinin B_2 receptors. J Exp Med 2000; 192:1289-1299.
38. Rodriguez A, Rioult MG, Ora A et al. A trypanosome-soluble factor induces IP_3 formation, intracellular Ca^{2+} mobilization and microfilament rearrangement in host cells. J Cell Biol 1995; 129:1263-1273.

39. Leite MF, Moyer MS, Andrews NW. Expression of the mammalian calcium signaling response to Trypanosoma cruzi in Xenopus laevis oocytes. Mol Biochem Parasitol 1998; 92:1-13.
40. Dorta ML, Ferreira AT, Oshiro MEM et al. Ca^{2+} signal induced by Trypanosoma cruzi metacyclic trypomastigote surface molecules implicated in mammalian cell invasion. Mol Biochem Parasitol 1995; 73:285-289.
41. Wilkowsky SE, Wainszelbaum MJ, Isola ELD. Trypanosoma cruzi: participation of intracellular Ca^{2+} during metacyclic trypomastigote-macrophage interaction. Biochem Biophys Res Commun 1996; 222:386-389.
42. Ruiz RC, Favoreto S, Dorta ML et al. Infectivity of Trypanosoma cruzi strains is associated with differential expression of surface glycoproteins with different Ca^{2+} signalling activity. Biochem J 1998; 330:505-511.
43. Tardieux I, Webster P, Ravesloot J et al. Lysosome recruitment and fusion are early events required for trypanosome invasion of mammalian cells. Cell 1992; 71:1117-1130.
44. Rodriguez A, Samoff E, Rioult MG et al. Host cell invasion by trypanosomes require lysosomes and microtubule/kinesin-mediated transport. J Cell Biol 1996; 134:349-362.
45. Andrews N. Lysosome recruitment during host cell invasion by Trypanosoma cruzi. Trends Cell Biol 1995; 5:133-137.
46. Rodriguez A, Webster P, Ortego J et al. Lysosomes behave as Ca^{2+}-regulated exocytic vesicles in fibroblasts and epithelial cells. J Cell Biol 1997; 137:93-104.
47. Rodriguez A, Martinez I, Chung A et al. cAMP regulates Ca^{2+}-dependent exocytosis of lysosomes and lysosome-mediated cell invasion by trypanosomes. J Biol Chem 1999; 274:16754-16759.
48. Kima PE, Burleigh B, Andrews NW. Surface-targeted lysosomal membrane glycoprotein-1 (LAMP-1) enhances lysosome exocytosis and cell invasion by Trypanosoma cruzi. Cellular Microbiol 2000; 2:477-486.
49. Schenkman S, Robbins ES, Nussenzweig V. Attachment of Trypanosoma cruzi to mammalian cells requires parasite energy, and invasion can be independent of the target cell cytoskeleton. Infect Immun 1991; 59:645-654.
50. Schenkman SRA, Mortara R. HeLa cells extend and internalize pseudopodia during active invasion by Trypanosoma cruzi trypomastigotes. J Cell Sci 1992; 101:895-905.
51. Procópio DO, Barros HC, Mortara RA. Actin-rich structures formed during the invasion of cultured cells by infective forms of Trypanosoma cruzi. Eur J Cell Biol 1999; 78:911-924.
52. Barbosa HS, Meirelles MNL. Evidence of participation of cytoskeleton of heart muscle cells during the invasion of Trypanosoma cruzi. Cell Struct Funct 1995; 20:275-284.
53. Carvalho TMU, Ferreira AG, Coimbra ES et al. Distribution of cytoskeletal structures and organelles of the host cell during evolution of the intracellular parasitism by Trypanosoma cruzi. J Submicrosc Cytol Pathol 1999; 31:325-333.
54. Docampo R, Moreno SNJ. The role of Ca^{2+} in the process of cell invasion by intracellular parasites. Parasitol Today 1996; 12:61-65.

CHAPTER 6

The Role of Nitric Oxide in the Pathogenesis of *Trypanosoma cruzi* Infection

Gislâine A. Martins, Mauro M. Teixeira and João S. Silva

Summary

This Chapter summarizes the roles of nitric oxide (NO) in mediating and controlling the effector immune response against *Trypanosoma cruzi* and in the pathology resulting from the infection. The killing of the trypomastigote form of the parasite is dependent on the production of NO which is catalyzed by an inducible NO synthase (iNOS). The cytokines IFN-γ and TNF-α produced during the acute phase of infection appear to play a major role in the induction of iNOS. However, other molecules, such as chemokines and platelet–activating factor (PAF), can also induce NO production and NO-dependent killing of *T. cruzi* by murine macrophages. On the other hand, TGF-β and IL-10, which are also produced during the infection, are negative regulators of NO production. In addition to mediating resistance against the infection, NO can also suppress the immune response to *T. cruzi* via the induction of apoptosis of T cells. Furthermore, there is now clear evidence to suggest that NO is involved in the pathogenesis of neuronal and myocardial dysfunction in experimental models and in patients.

Nitric Oxide

Nitric oxide (NO) is a very diffusible gas generated from the oxidation of L-arginine to L-citrulline by a family of NADPH-dependent enzymes (nitric oxide synthases, or NOS). The family of NO-generating enzymes consists of three isoenzymes: a constitutively expressed neuronal NOS (bNOS or NOS1), an endothelial NOS (eNOS or NOS3) and an inducible NOS (iNOS or NOS2). These enzymes are homodimers whose monomers are themselves two enzymes fused together—a cytochrome reductase and a cytochrome that requires three cosubstrates (L-arginine, NADPH and O_2) and five co-factors or prosthetic groups (FAD, FMN, calmodulin, tetrahydrobibiopterin and heme). The constitutively expressed NOS depends on intracellular Ca^{2+} levels to be active and leads to the production of low amounts of NO. In contrast, the inducible form of NOS is Ca^{2+} independent and when induced by diverse stimuli, such as microbial and/or cytokines, is able to generate far higher and enduring NO levels. Many cell types including macrophages, muscle cells, hepatocytes, fibroblasts, astrocytes and endothelial cells express iNOS (reviewed in refs. 1-3). However, there is a marked variability in the expression of iNOS and the production of NO in different tissues and different species.[4-6] For example, stimuli known to readily induce iNOS expression in murine tissue macrophages do not induce iNOS expression in human mononuclear phagocytes purified from healthy human blood. However, iNOS expression can be induced in human macrophages by alternative stimulatory mechanisms, such as culturing in presence of anti-IgE receptor (CD23).[7] Moreover,

Molecular Mechanisms in the Pathogenesis of Chagas Disease, edited by John M. Kelly.

iNOS expression can be found in human tissue macrophages and, to a lesser extent, in blood monocytes after infection with certain pathogens.[8-10]

As a very reactive compound, NO is potentially able to react with the redox forms of oxygen, thiols, amines and transition metals (reviewed in refs. 3,11). A common biochemical trigger utilized in NO reactions is the S-nitrosylation or nitration of proteins. This property enables NO to be involved in many biological functions. As a neurotransmitter, in very low concentrations (c-NOS-derived), NO regulates intestinal peristalsis, autonomic and neuroendocrine functions. As a cytotoxic/cytostatic effector molecule, when produced in higher amounts (iNOS-derived), NO is potentially microbicidal. Moreover, depending on its concentration, the biological redox milieu and the involvement/induction of intracellular mechanisms, NO can interfere with cell proliferation and death by either inducing or suppressing apoptosis.[12-13]

As a microbicidal agent, NO has been shown to play a substantial role in protective immunity against many pathogens, including *Mycobacterium tuberculosis,*[14] *Salmonella typhimurium,*[15] *Listeria monocytogenes,*[16] *Leishmania major,*[17-18] *Leishmania donovani,*[19] *Trypanosoma cruzi,*[20-21] *Toxoplasma gondii*[22] and Coxsackie B3 virus,[23] among others (See Table 1). However, the influence of NO in immune mechanisms extends beyond its direct antiparasitic effects and it has been suggested that it may play a prominent immunoregulatory role in different disease models. In this Chapter, we have aimed to provide an overview and to discuss the participation of NO in the killing, pathology and the immune response associated with infection by *T. cruzi*.

NO and Parasite Killing

The antimicrobial activity of reactive nitrogen intermediates (RNIs) and especially of NO has been most convincingly demonstrated by the use of mice deficient in iNOS (see Table 1). These studies show that a broad spectrum of infectious agents (ranging from viruses to helminths) are directly or indirectly controlled by RNIs in vivo (reviewed in refs. 6,24,25). The dependence of NO biosynthesis on the mechanisms that control intracellular multiplication of *T. cruzi* in vivo and in vitro has been broadly demonstrated. Results from our own and other laboratories have clearly shown a role for macrophage-derived NO in the control of *T. cruzi* infection in mice.[26-28,20-21] iNOS synthesis in macrophages is driven by IFN-γ and TNF-α (reviewed in ref. 8) and NO production in mice, as assessed by plasma levels of nitrate, reaches maximal levels around 8-12 days post-infection with *T. cruzi.*[27] The main initial source of IFN-γ appears to be NK cells stimulated by IL-12 and TNF-α (Fig. 1).[28-29] On the other hand, cytokines such as TGF-β and IL-10 are frequently implicated as negative regulators of NO production by activated macrophages (reviewed in ref. 8), especially *T. cruzi*-infected macrophages.[26] The modulatory effects of IL-10 during *T. cruzi* infection is still a controversial issue: It was firstly demonstrated that production of IL-10 was significantly increased in susceptible mice as compared to the resistant ones.[30] Next, Reed and collaborators[31] showed that in a susceptible mouse strain, administration of neutralizing anti-IL-10 antibodies confers resistance to the infection and that this is related to increased levels of IL-12 and IFN-γ and possibly of NO. However, the same treatment did not confer protection when performed in 129/SvEv mice infected with the Y strain[32] of *T. cruzi*. In addition, IL-10 deficient mice (IL-10-/-) presented higher parasitemia and similar mortality rates when compared to wild type mice infected with the Y strain. Finally, a more recent study has shown that when infected with the Tulahuem strain of *T. cruzi*, IL-10-/- mice present enhanced mortality rates but decreased parasitemia levels associated with increased production of Th1 cytokines.[33] This is a possible indirect indication of increased NO production.

Similarly, administration of TGF-β to mice infected with *T. cruzi* leads to increased parasitemia and mortality, which is associated with decreased production of IFN-γ, in vivo and in vitro.[34] In addition, its has been proposed that TGF-β produced by macrophages during the

Table 1. Disease severity in iNOS (NOS2) knock out mice

Increased Disease Severity		**Unchanged** Disease Severity	
Infectious Agent/Route of Infection	References	Infectious Agent/Route of Infection	References
Herpes simplex virus type 1 (cutaneous infection)	104	*Chlamydia trachomatis* (vaginal infection)	112-113
Murine cytomegalovirus	105	*Helicobacter pylori*	114
Coxsackie virus B3 (i.p.)	23	*Salmonella typhimurium* (i.p.)	15
Ectromelia virus (i.v.)	106	*Shigella flexneri* (i.n.)	115
Entamoeba histolytica	107	*Trypanosoma brucei* (i.p.)	116
Staphilococcus aureus (i.v.)	108	*Plasmodium berghei* (i.v.)	117-118
Chlamydia pneumoniae (vaginal infection)	109	*Plasmodium chabaudi*	119
		Borrelia burgdorferi (s.c.)	120
Listeria monocytogenes	16		
Mycobacterium tuberculosis (i.v.)	14		
Klebsiella pneumonia	110		
Salmonella typhimurium i.p.	15		
Leishmania major (s.c.)	17-18		
Leishmania donovani (i.v.)	19		
Toxoplasma gondii (i.p. or p.o.)	22		
Trypanosoma cruzi (i.p.)	20-21		
Trypanosoma brucei (i.p.)	111		
Schistosoma mansoni #	73		

In this study mice were previously vaccinated and then infected with *S. mansoni*. The protective effect of vaccination was reduced in iNOS-/- mice, as compared with wild type mice.

acute phase of infection could contribute to parasite growth by increasing macrophage metabolism via arginase,[35] which produces ornithine and urea, required for parasite survival.[36] The evidence of a role for NO in the control of experimental *T. cruzi* infection derives from studies evaluating the effects of NO synthesis inhibitors, which limit the expression of iNOS and studies in iNOS-deficient animals. Thus, treatment of mice with L-arginine analogs, such as N^G- methyl-L-arginine (L-NMMA),[27] aminoguanidine (AG) (our unpublished results) or L iminoethyl-L-ornithine (L-NIL)[37] leads to increased parasitemia and mortality in mice infected with *T. cruzi*. Similarly, pretreatment with monoclonal antibodies against the NO-inducing cytokines IFN-γ or TNF-α prevented iNOS expression and NO production, resulting in greater parasitemia and mortality.[38] In agreement with an important role of NO in controlling *T. cruzi*, iNOS-/- infected mice are highly susceptible to infection with at least three different strains of *T. cruzi* (Fig. 2).[20-21] Moreover, the NO donor drug S-nitroso-acetyl-penicillamine (SNAP) has been shown to kill *T. cruzi* trypomastigote forms in vitro in the absence of any other cells, indicating a direct NO-mediated killing of this parasite.[27]

A few studies have attempted to elucidate some of the molecular mechanisms by which NO mediates its cytotoxic effects against *T. cruzi*. Venturini and collaborators[39] have shown that NO efficiently inhibits the activity of cruzipain, a major cysteine proteinase expressed in all life-cycle stages of the parasite and which is most abundant in the replicating forms (for a review see ref. 40). Cruzipain has an important role in parasite nutrition, is implicated in

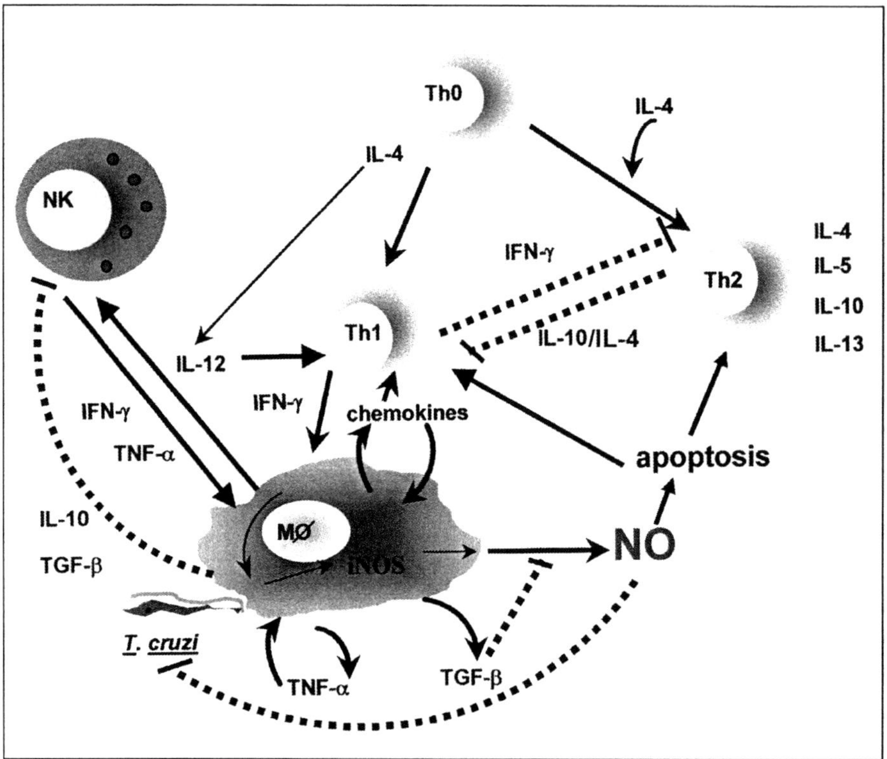

Figure 1. Regulation of NO production by cytokines and chemokines in *T. cruzi*-infected macrophages. Infection of macrophages with *T. cruzi* can result in chemokine,[45] and IL-12[94] production (solid lines indicate induction/activation). IL-12 induces production of IFN-γ[29] and TNF-α by NK cells. In addition, IL-12 favors differentiation of Th1cells, which will result in additional IFN-γ production. Together, IFN-γ and TNF-α induce iNOS activity and NO production by macrophages[28] which in turn mediates parasite killing (broken lines indicate inhibition) but also modulates the immune response by inducing apoptosis.[38] On another hand, infection of macrophages with *T. cruzi* could also result in production of TGF-β,[34] which down regulates macrophage activation and NO production. NO production can also be indirectly down regulated by IL-10 produced by Th2 differentiated T cells.

modulating cell invasion and also participates in the mechanisms used by the parasite to escape immune attrition (reviewed in ref. 41). As such, NO-mediated inactivation of this protein could represent an important trypanocidal pathway. The rapid generation of peroxynitrite anion (ONOO-) from NO and superoxide, which is produced simultaneously in macrophages and other leukocytes, may also play a role in the trypanocydal activity of NO. Peroxynitrite kills *T. cruzi* in a dose-dependent manner[42] by a mechanism probably involving inhibition of calcium uptake by the parasites.[43]

In addition to the fundamental role of IFN-γ and TNF-α in synergising to induce NO production by *T. cruzi*-infected macrophages, there is now clear evidence that mediators which activate cells by interacting with seven-transmembrane G protein-coupled receptors can also induce NO release from macrophages. Amongst such mediators, a role for platelet activating factor (PAF) and chemokines has been demonstrated by our group.[44-45] PAF activates infected, but not naïve, macrophages to release significant amounts of NO which is sufficient for the killing of the intracellular parasites.[44] The effects of PAF were receptor mediated and could be

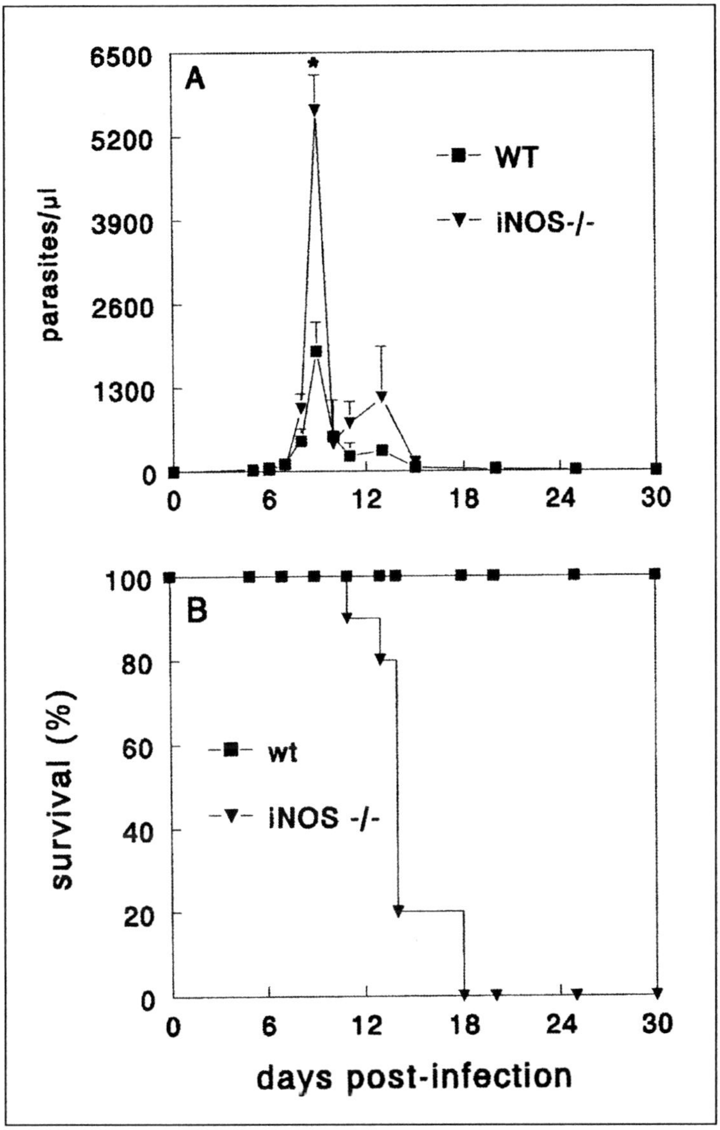

Figure 2. iNOS-deficient mice present increased parasitemia and mortality after infection with *T. cruzi*. WT control (squares) (n=10), and iNOS-deficient (iNOS-/-) (n=15) mice (triangles) were each infected i.p. with 10^3 blood trypomastigotes forms (Y strain), and the parasitemia (n= 10), (A) and mortality (B) were evaluated. The results in panels A and B are expressed as median ± SEM. Data from two experiments are shown. The asterisk in A indictates where the difference between WT and iNOS-/- mice has statistical significance, p<0.01 (Student-Newman-Keuls test).

blocked by pertussis toxin, which disrupts G protein signaling. Interestingly, the ability of PAF to induce NO was blocked by anti-TNF-α, suggesting that the production of low levels of TNF-α was necessary for the induction of iNOS. The ability of PAF to induce NO in vivo has been shown to be important by experiments which evaluated the effects of treating mice with

PAF receptor antagonists during the course of *T. cruzi* infection. Not only were parasite levels markedly increased in treated animals, but there was also significantly greater mortality than in the control group.[44] Similarly, PAF induced NO production and NO-dependent killing of *T. cruzi* by rat macrophages and treatment of *T. cruzi*-infected rats with a PAF receptor antagonist was followed by a significant increase in tissue parasitemia (our unpublished results).

More recently, we and others have shown that chemokines, mediators which also function by activating seven-transmembrane G protein-coupled receptors,[46] can also induce NO production and NO-dependent killing of *T. cruzi* by murine macrophages.[45,47] Of particular interest, there is appreciable expression of chemokines in infected tissues during the course of experimental *T. cruzi* infection in mice[48-49] and also in *T. cruzi*-infected IFN-γ-primed macrophages, especially MCP-1 and RANTES.[45] Moreover, chemokines seem to synergize with low doses of IFN-γ to induce further NO release and parasite killing. Overall, these studies suggest that the activation of serpentine G protein-coupled receptors is an important mechanism for inducing NO production by infected macrophages and that this mechanism may be operative in vivo, especially in the presence of IFN-γ.

Much less is known about the mechanism that underlies the ability of the chemokines to induce NO release by infected macrophages. One possible explanation came from observations on the number of parasites inside macrophages early in the course of infection in vitro.[44-45] In these studies, we showed that the number of parasites inside cells at 48 h is markedly diminished compared with the number at 4 h, reflecting trypanocidal activity. Thus, it appears that chemokines and PAF are facilitating the ability of macrophages to phagocytose *T. cruzi* trypomastigotes, which is apparently enough to trigger the activation of iNOS. In support of this hypothesis, pre-treatment of macrophages with drugs which disrupt phagocytosis (i.e., cytochalasins) is followed by a marked drop in the uptake of parasites early in the course of infection and a decrease in NO production (our unpublished data).

Parasite-Derived Products Induce NO Production

Bacterial lipopolysaccharide (LPS) was one of the first microbial products shown to activate leukocytes to produce a range of inflammatory cytokines and NO, especially in the presence of IFN-γ. Host cells can recognize not only LPS, but also a range of microbial-specific molecules present in bacteria, mycobacteria and protozoan parasites. These parasite-associated molecular patterns (PAMPs) present in microbes, but not in host cells, are recognized by specialized receptors called Toll-like receptors (TLR) that transduce information to the nucleus.[50-51] Akin to bacterial LPS and mycoplasma LPG, *T. cruzi* trypomastigotes express glycosylphosphatidylinositol (GPI)-anchored mucin-like glycoproteins (tGPI-mucins) (Fig. 3) on their surface which are capable of activating IFN-γ-primed murine macrophages to induce NO and the production of pro-inflammatory cytokines, such as TNF-α (see Fig. 4) and IL-12.[52-54] It appears that the GPI-anchors are the specific part of the molecule that is capable of activating macrophages.[55] Pre-treatment of macrophages with IL-10 or drugs which elevate cyclic AMP modulate the ability of tGPI-mucins to induce NO and pro-inflammatory cytokines production by macrophages.[56] The immunomodulatory role of IL-10 on TNF-α and NO release is also observed in vivo (see above). In contrast to LPS, but similar to gram-positive bacteria-derived lipoteicoic acid, tGPI mucins activate the TLR-2 receptor in macrophages to induce NO, TNF-α and IL-12 production.[57]

Thus, a scenario emerges where not only can the parasite itself, but also parasite-derived products, activate host immune cells. The ability of host cells to respond to parasite molecules may be advantageous as a means of non-specific microbe detection that could facilitate a rapid non-specific immune response to deal with the infection. However, in the absence of live parasites, parasite-derived products may activate immune cells and potentially lead to undesirable inflammation and damage to tissue cells. In this regard, LPS is thought to play a major role in

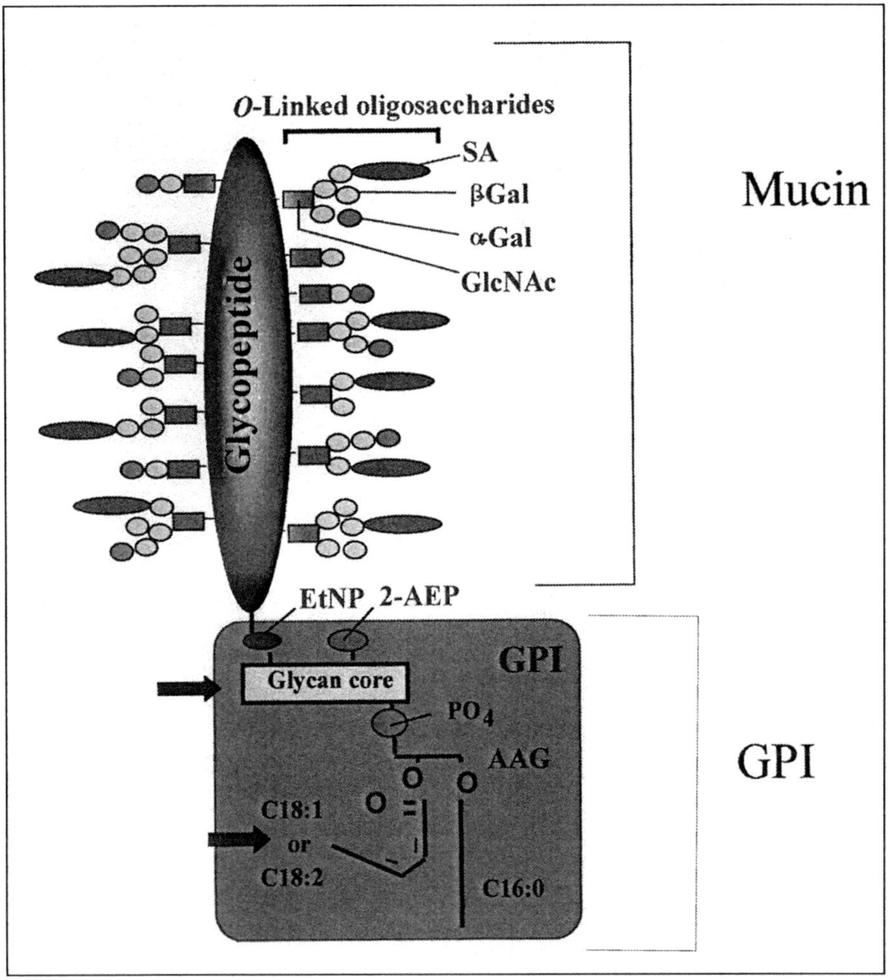

Figure 3. Schematic structure of *T. cruzi* trypomastigote mucins. The GPI anchor is the active part of the molecules and is shown in the gray square. Black arrows indicate the glycan core (top) and fatty acids (bottom), two portions implicated in the activation of macrophages by *T. cruzi* mucins.

the systemic inflammatory syndrome that accompanies bacterial infection in humans. Whether tGPI-mucins also play a similar important pathophysiological role in Chagas disease is not known but clearly warrants further investigation. Of note is the fact that tGPI-mucins can induce the production of the chemokine macrophage chemotactic protein-1 (MCP-1) by macrophages both in vitro and in vivo and that the MCP-1 produced is capable of inducing the infiltration of mononuclear cells in vivo (our own unpublished data).

NO As an Immunomodulator

The first indication that NO could play a role in the regulation of immune responses came from the observation that the high concentration of NO produced by rat spleen cells in response to some stimuli, such as phytohaemagglutinin (PHA) or during mixed lymphocyte responses (MRL), was able to inhibit the proliferation of T cells.[58] These findings have been

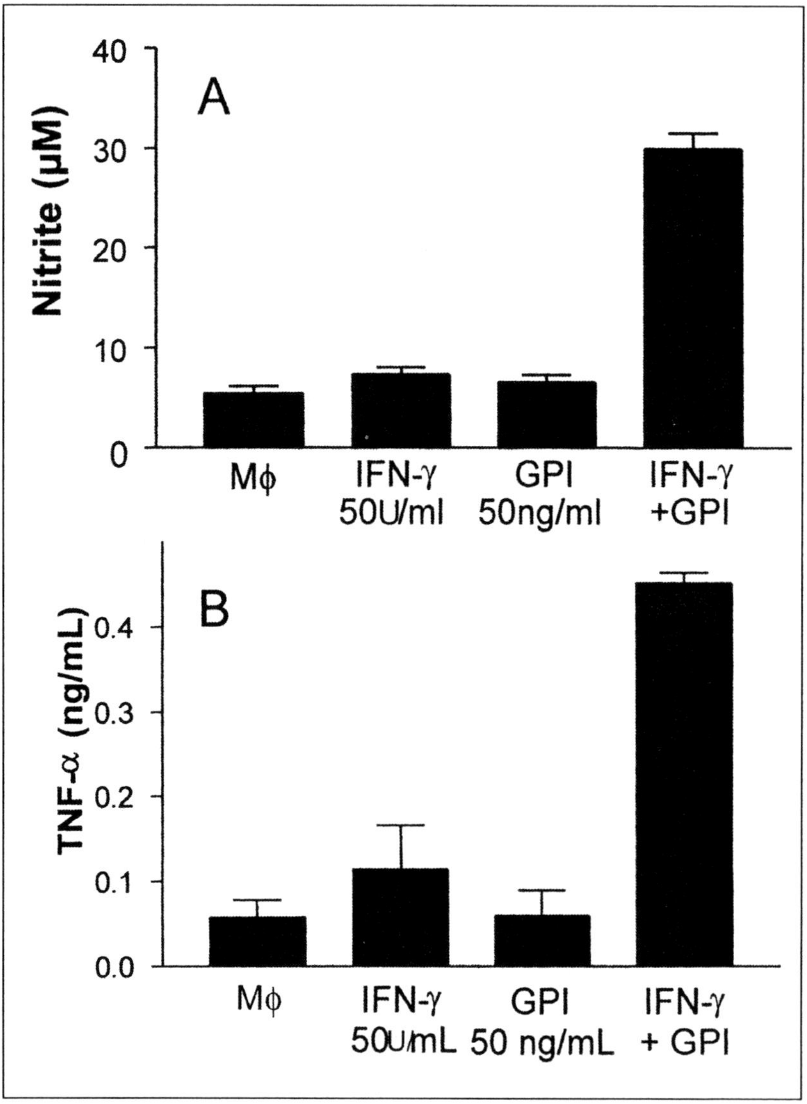

Figure 4. GPI synergizes with IFN-γ to induce NO and TNF-α production. Peritoneal murine macrophages were cultured in medium only or in presence of GPI (50 ng/ml), recombinant murine IFN-γ (50 ng/ml) or GPI plus IFN-γ. After 48 hours of culture, the concentration of nitrite (A) and TNF-α (B) were determined using Griess reaction and TNF-α specific ELISA, respectively.

extended by the demonstration that NO, resulting from activated macrophages in mice infected with many different parasites, including *Toxoplasma gondii,*[59-60] *Trypanosoma brucei,*[61] *Plasmodium chabaudi chabaudi*[62] and *Listeria monocytogenes*[63] among others, exerts potent T cell suppressive activity. In mice infected with any of these pathogens the production of NO is generally increased soon after infection and acts to suppress mitogen and specific antigen-induced T cell proliferation. This can be partially restored by the addition of inhibitors of RNI production. The suppressive effects of NO on the lymphoproliferative response to parasite antigens and to polyclonal T cell stimuli have also been described in mice infected with *T. cruzi*.[32] Inhibition of NO production by addition of a L-arginine analog, L-NMMA, or inhibition of iNOS expression with antibodies against IFN-γ or TNF-α resulted in partial restoration of the proliferative response in infected mice.[32,38]

Although the molecular mechanisms underlying the suppressor role of RNIs are not completely understood, it has now become clear that NO is able to interfere with signaling pathways related to cell proliferation. For example, NO can prevent the activation of Janus Kinase[64] and inactivate enzymes required for DNA synthesis, such as ribonucleotidyl reductase.[65] In addition, NO can inactivate enzymes necessary for mitochondrial function and cell metabolism,[66] modulate NFκB activation[67] and suppress IL-2 gene expression at the level of transcription, by interfering with zinc-finger transcription factor activity.[68]

As a very reactive molecule that can interfere with the cell cycle and is able to react with a full range of other biological molecules, NO also appears to modulate the pathways leading to apoptosis. Depending on its concentration and on the redox state of the cell, NO can promote or protect cells from apoptosis by interfering with many pathways, including the modulation of caspase activity (reviewed in refs. 12 and 69). Considering that apoptosis appears to be an important mechanism that regulates T-cell expansion during immune responses, it can be envisaged that NO potentially regulates immune responses, possibly by protecting and/or inducing apoptosis in activated T cells. In this regard, our previous studies have indicated that the induction of apoptosis is one of the mechanisms by which NO suppresses T cell responses in mice acutely infected with *T. cruzi*.[38] Consistent with this, the absence of iNOS activity results in reduced apoptosis in mice acutely infected with *T. cruzi*.[21,38] As a consequence, NO might play a role in the control of inflammatory processes or in deletion of autoreactive T cells, thereby fulfilling host-protection functions during the immune response against parasites (reviewed in refs. 24,70).

A broad ability to modulate cytokine production has been attributed to NO. It has been reported that it can affect the production of more than 20 cytokines, including IL-1, IL-6, IL-4, IL-8, IL-10, IL-12, IFN-γ, TNF-α and TGF-β (reviewed in ref. 24). Of great interest in the context of infectious disease, NO has been shown to mediate specific impairment of T helper 1 (Th1) cells with concomitant suppression of IL-2 and IFN-γ production, while Th2-cell function appears largely unaffected.[71-72] Thus, iNOS knock out mice develop a significantly stronger Th1 response than the wild-type when infected with *L. major*[18] or with *Schistosoma mansoni*.[73] Moreover, results from our laboratory have shown that iNOS-/- infected mice produce significantly more IFN-γ than wild type infected mice.[21] In agreement, it has been reported that IFN-γ production is increased in iNOS-deficient mice infected with other pathogens, such as *Mycobacterium*[14] and influenza A virus.[74] Furthermore, at least in the human immune system, NO can promote IL-4 production and, as a consequence, limit Th1-cell activity.[75]

Besides the direct interference with production of cytokines by T cells, NO can induce transcription of the IL-12 p40 gene, but not of the p35 gene in macrophages.[76] Because the IL-12 $(p40)_2$ homodimer is an antagonist of IL-12,[77] the overproduction of p40 may contribute to the down-regulation of Th1 reactivity in the presence of NO. In addition, in antigen presenting cells (APC), NO can inhibit major histocompatibility complex (MHC) class II expression[79] On the other hand, iNOS activity seems to be required to achieve full IL-12 mediated-activation via STAT-4 phosphorylation in NK cells during the innate response to a protozoan parasite.[78] These studies suggest that NO could be modulating the production of IFN-γ during innate immunity, which could favor or antagonize the development of Th1 later in the immune response. In addition, NO seems to increase production of prostaglandin E_2 (PGE_2) in murine macrophages.[79]

NO and Pathology

The pathogenesis of chronic chagasic cardiopathy is still a controversial issue. The scarcity of *T. cruzi* parasites in the myocardium and the supposed lack of correlation between parasite presence and the occurrence of myocardial inflammatory infiltrate has led to theories that chronic Chagas cardiopathy is an autoimmune disease. The $CD4^+$ T cell-mediated rejection of syngeneic heart grafts by mice chronically infected with *T. cruzi* had suggested that autoimmunity could be the major mechanism implicated in the pathogenesis of chagasic cardiomyopathy in humans.[80] However, more recent studies using highly sensitive methodology have demonstrated the existence of a strong association between moderate or severe myocarditis and the presence of *T. cruzi* antigens in the heart, suggesting the direct participation of parasites in the genesis of the pathology associated with the chronic disease.[81] In fact, *T. cruzi* antigens are frequently detected in chronic Chagas heart disease, and are associated with an intense inflammatory infiltrate.[82] Together these data provided strong evidence for the role of *T. cruzi* in triggering the inflammatory reaction in the heart.

The inflammatory cell infiltrate which characterizes human chronic chagasic myocarditis is composed mainly by $CD4^+$ and $CD8^+$ T cells. The presence of the latter subset seems to correlate directly with the presence of parasite antigens.[83] Furthermore, cytokines such as IFN-γ, TNF-α and IL-1 can be detected in the inflammatory cell infiltrate in both human and experimentally-infected animals.[84-85] Interestingly, the presence of these inflammatory cytokines in the myocardium was shown to correlate with iNOS activation and NO production.[86] In fact, results from our and other laboratories have shown that iNOS activity is abundant in the heart of *T. cruzi*-infected mice[87-88] and of chagasic patients (V. Rodrigues, unpublished results).

In the last few years, NO has been found to be involved in the pathogenesis of myocardial dysfunction associated with a variety of cardiomyopathies and myocarditis. NO-associated changes in myocardial contractility may be an important factor in several disease states, including septic shock, rejection of heart transplants and viral induced myocarditis. The mechanism of contractile depression by NO is only partially understood, but NO is able to decrease the intracellular levels of cAMP in response to beta-adrenergic stimulation in cardiac myocytes, at least in part through a cGMP-mediated mechanism.[89] Moreover, macrophage derived cytokines also exert negative inotropic and toxic effects on cardiac myocytes through the transcriptional activation of iNOS and subsequent production of NO. Time-course studies have implicated iNOS-derived NO as a mediator of cardiac myocyte apoptosis.[90] In fact, high levels of NO production by iNOS is able to kill cardiac myocytes by triggering apoptosis, possibly by a p53-mediated mechanism.[91] iNOS activation has been found in the heart of chagasic patients and of *T. cruzi* infected mice and NO has been implicated as a mediator of apoptosis in immune cells in the periphery (as discussed above). However, studies performed in situ in preserved heart tissue from chagasic patients were unable to detect a significant increase in DNA fragmentation and p53 expression in myocadial cells or mononuclear cells.[92]

Expression of IFN-γ, TNF-α, IL-1β and iNOS mRNA can be detected in heart tissue of *T. cruzi*-infected mice or rats.[85,88] In addition, isolated fetal murine cardiomyocytes cultured in presence of trypomastigote forms of *T. cruzi* express mRNA for the cytokines TNF-α and IL-1β and for iNOS, strongly suggesting that these cells could be the potential sources of cytokines and iNOS in vivo. Moreover, following infection of cultured myocytes with *T. cruzi* we found iNOS protein induction and NO-production, which could be blocked by selective iNOS inhibitors (L-NIO and aminoguanidine), demonstrating that the parasites induced NO production in cardiomyocytes via upregulation of the expression of the inducible isoform of NOS.[88]

The mechanism by which iNOS expression is induced by *T. cruzi* in cardiac myocytes remains unresolved. One possibility is that parasite-secreted products, such as tGPI mucins[52] or LPS-like molecules,[93] may induce the enzyme directly. Alternatively, iNOS expression may result from autocrine stimulation by cytokines and chemokines released by cardiomyocytes following *T. cruzi* infection. In this regard, *T. cruzi* has been shown to induce production of β-chemokines by macrophages[45] and the expression of JE, RANTES, KC, MIP-2, Mig and Crg2 mRNA in cardiomyocytes.[48,88] The parasites can also induce TNF-α and IL-12 synthesis by macrophages, which results in IFN-γ production by T and NK cells.[28,94] The presence of IFN-γ and chemokines in the heart tissue of infected mice, in association with IL-1β and TNF-α could lead to induction of iNOS. Besides its potential toxic effects, NO could also participates in the control of parasite replication in the heart. In fact, incubation of cardiomyocytes with cytokines or chemokines resulted not only in NO synthesis but also in significant trypanocidal activity. Addition of selective iNOS inhibitors significantly inhibited NO production and parasite killing, convincingly demonstrating that cardiomyocyte-derived NO possesses significant trypanocidal activity.[88]

The importance of the production of chemotactic cytokines by cardiomyocytes for disease outcome and host immunopathology during *T. cruzi* infection is not completely understood, but these cells may be playing an important role in driving or maintaining the local inflammatory response. It is possible that the early chemokine-and cytokine-mediated cardiomyocyte activation could play an essential role in the containment of parasite dissemination during the acute phase of infection. On the other hand, the release of parasites from the amastigote nests within cardiac tissue may act locally to enhance the response which controls parasite replication and spread in the host tissues (see Fig. 5). Nevertheless, the induction of these molecules in cardiomyocytes could also play a detrimental role in chronic Chagas heart disease.[95-96] Cytokines such as TNF-α, for instance, may be autocrine contributors to the myocardial dysfunction seen in chagasic patients.[96] Moreover, NO may directly regulate the contractile properties of muscle cells,[97] leading to depressed cardiac function[89] and myocardial damage.[98] This could further potentiate chemokine-driven inflammation and tissue damage.

On the other hand, the finding that *T. cruzi*-infected and L-NMMA-treated mice have an increased severity of lesions in skeletal muscle and liver[99] together with the observation that iNOS-/- mice have increased myocardial inflammation and a different pattern of chemokine production when infected with the Colombian strain of *T. cruzi* (our unpublished results) could suggest that NO might be broadly implicated in modulating inflammatory responses through modulation of Th1 cytokine production in the acute phase of the infection (Fig. 5). In this regard, one could presume that Th1 cells (or IFN-γ producers) would predominate in inflammatory lesions in iNOS-/- infected mice as compared to the WT infected mice. This is a tempting possibility, but still remains to be investigated. Nevertheless, the enhanced inflammatory activity in the absence of iNOS could be related to the interference that NO can exert on cell migration. As recently reported, NO is able to inhibit leukocyte adhesion and migration through the endothelial cell layer by downregulating expression of selectins, vascular cell adhesion molecule (VCAM) and intracellular adhesion molecule 1 (ICAM-1).[100-101] Furthermore, P-selectin expression was found to be impaired in the presence of NO.[102] Since P- and E-selectins mediate recruitment of Th1 but not Th2 cells into inflamed tissues[103] it is conceivable that NO could preferentially downregulate the accumulation of Th1 cells at the sites of chronic inflammation by interfering with the adhesion process.

Concluding Remarks

Substantial progress has been made in understanding the important role played by NO in modulating several biological phenomena, including its crucial importance as a microbicidal agent. However the mechanisms by which NO modulates the immune response are just beginning to be elucidated. It is now clear that the induction of NO production during *T. cruzi* infection is an important microbicidal mechanism. However, NO can certainly play additional roles during the immune response against the parasite, due to its ability to modulate cytokine production, cell migration and apoptosis. There are many questions to be answered about the

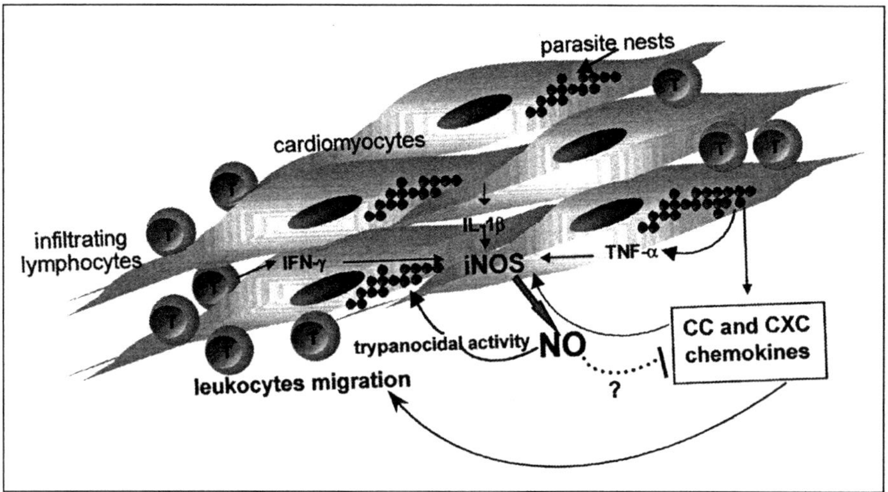

Figure 5. Modulation of chemokine and NO production by cardiomyocytes infected with *T. cruzi*. Infection of cardiomyocytes with *T. cruzi* leads to the production of TNF-α, IL-1β and the MIG, KC, JE/MCP-1, MIP-2 and Crg-2 chemokines,[88] which potentiates iNOS activity and NO synthesis.[88] Besides mediating potent trypanocidal activity, NO appears to downmodulate chemokine production by infected cardiomyocytes. In vivo, the chemokines produced by *T. cruzi*-infected cardiomyocytes can also induce leukocyte migration resulting in IFN-γ production and a maintained IL-1β and TNF-α secretion at the inflammatory site. These cytokines, in turn, lead to increased chemokine production by infected cardiomyocytes.

exact contribution of NO to the control of immune responses, especially during the establishment of chronic inflammatory processes such as the *T. cruzi*-induced cardiomyopathy. Greater understanding of the interaction between chemokines, cytokines and NO during *T. cruzi* infection may have important implications for the development of future therapies designed to protect the host against the pathology induced by the parasite.

References

1. Stamler JS, Singel DJ, Loscalzo J. Biochemistry of nitric oxide and its redox-activated forms. Science 1992; 258:1898-1902.
2. Stuehr D. Mammalian nitric oxide synthases. Biochim Biophys Acta 1999; 1411:217-230.
3. Nicotera P, Bernassola F, Melino G. Nitric oxide (NO), a signalling molecule with a killer soul. Cell Death Differ 1999; 4:435-442.
4. Zhang X Laubach V, Alley E et al. Transcriptional basis for hyporesponsiveness of the human inducible nitric oxide syntase gene to lipopolysaccaride/interferon-gamma. J Leuk Biol 1996; 59:575-585.
5. Rothenberg ME, Örn A. Conflicting roles of NO in Trypanosoma cruzi infection. The Immunologist 1997; 5/4:127-132.
6. Nathan C, Shiloh UM. Reactive oxygen and nitrogen intermediates in the relationship between mammalian hosts and microbial pathogens. Proc Natl. Acad Sci USA 2000; 97:8841-8848.
7. Vouldoukis I, Rivieros-Moreno D, Dugas B et al. The killing of Leishmania major by human macrophages is mediated by nitric oxide induced after ligation of the Fc epsilon RII/CD23 surface antigen. Proc Natl Acad Sci USA 1995; 92:7804-7808.
8. MacMincking J, Xie QW, Nathan C. Nitric oxide and macrophage function. Annu Rev Immunol 1997a; 15:323-350.

9. Weinberg JB. Nitric oxide production and nitric oxide syntase type 2 expression by human mononuclear phagocytes: A review. Mol Med 1998; 4:557-591.
10. Facchetti F, Vermi W, Fiorentini S et al. Expression of inducible nitric oxide synthase in human granulomas and histiocytic reactions. Am J Pathol 1999; 154:145-152.
11. Stamler JS. Redox signaling: Nitrosylation and related target interactions of nitric oxide. Cell 1994; 78:931-936.
12. Brune B, von Knethen A, Sandau K. Nitric oxide: An effector of apoptosis. Cell Death Diff 1999; 6:969-975.
13. Liu L, Stamler JS. NO: An inhibitor of cell death. Cell Death Diff 1999; 6: 937-942.
14. MacMicking JD, North RJ, LaCourse R et al. Identification of nitric oxide synthase as a protective locus against tuberculosis. Proc Natl Acad Sci USA 1997b; 94:5243-5248.
15. Shiloh MU, MacMicking JD, Nicholson S et al. Phenotype of mice and macrophages deficient in both phagocyte oxidase and inducible nitric oxide synthase. Immunity 1999; 10:29-38.
16. MacMicking JD, Nathan C, Hom G et al. Altered responses to bacterial infection and endotoxic shock in mice lacking inducible nitric oxide synthase. Cell 1995; 81:641-650.
17. Diefenbach A, Schindler H, Donhauser N et al. Type 1 interferon (IFN α/β) and type 2 nitric oxide synthase regulate the innate immune response to a protozoan parasite. Immunity 1998; 8:77-87.
18. Wei XQ, Charles IG, Smith A et al. Alterated immune response in mice lacking inducible nitric oxide synthase. Nature 1995; 375: 408-411.
19. Murray HW, and Nathan CF. Macrophage microbiocidal mechanisms in vivo: reactive nitrogen vs. Oxygen intermediates in the killing of intracellular Leishmania donovani. J. Exp. Med 1999; 189:741-746.
20. Holscher C, Kohler G, Muller U et al. Defective nitric oxide effector functions lead to extreme susceptibility of Trypanosoma cruzi-infected mice deficient in gamma interferon receptor or inducible nitric oxide synthase. Infect. Immun 1998; 66:1208-1215.
21. Martins GA, Petkova SB, Machado FS et al. Fas-FasL interaction modulates nitric oxide production in Trypanosoma cruzi-infected mice. Immunology 2001; 103:122-129.
22. Shartton-Kersten TM, Yap G, Magram J et al. Inducible nitric oxide is essential for host control of persistent but not acute infection with the intracellular pathogen Toxoplasma gondii. J Exp Med 1997; 185:1261-1273.
23. Zaragoza C, Ocampo C, Saura M et al. The role of inducible nitric oxide synthase in the host response to Coxsackievirus myocarditis. Proc Natl Acad Sci USA 1998; 95:2469-2474.
24. Bogdan C, Röllinghoff M, and Diefenbach A. Reactive oxygen and reactive nitrogen intermediates in innate and specific immunity. Curr Op Immunol 2000; 12:64-76.
25. Bogdan C. Nitric oxide and the immune response. Nat Immunol 2001; 10:907-916.
26. Gazzinelli RT, Oswald IP, Hieny S et al. The microbicidal activity of interferon-γ treated macrophages against Trypanosoma cruzi involves an L-arginine-dependent, nitrogen oxide-mediated mechanism inhibitable by interleukin-10 and transforming growth factor-β. Eur. J. Immunol. 1992; 22:2501-2506.
27. Vespa GNR, Cunha FQ and Silva JS. Nitric oxide is involved in control of Trypanosoma cruzi-induced parasitemia and directly kills the parasite in vitro. Infect. Immun 1994; 62:5177-5182.
28. Silva JS, Vespa, GNR, Cardoso MAG et al. Tumor necrosis factor alpha mediates resistance to Trypanosoma cruzi infection in mice by inducing nitric oxide production in infected gamma interferon-activated macrophages. Infect. Immun. 1995; 63:4862-4867.
29. Cardillo F, Voltarelli JC, Reed SG et al. Regulation of Trypanosoma cruzi infection in mice by gamma interferon and interleukin 10: Role of NK cells. Infect. Immun 1995; 64:128-134.
30. Silva JS, Morrissey PJ, Grabstein KH et al. Interleukin 10 and interferon gamma regulation of experimental Trypanosoma cruzi infection. J Exp Med 1992; 175:169-174.
31. Reed SG, Brownell CE, Russo DM et al. IL-10 mediates susceptibility to Trypanosoma cruzi infection. J Immunol 1994; 153:3135-3140.
32. Abrahamsohn, IA, Coffman RL. Cytokine and nitric oxide regulation of the immunossupression in Trypanosoma cruzi infection. J Immunol 1995; 155:3955-3963.
33. Hunter CA, Ellis-Neyes LA, Slifer T et al. IL-10 is required to prevent immune hyperactivity during infection with Trypanosoma cruzi. J Immunol 1997; 158:3311-3316.

34. Silva JS, Twardzik DR, Reed SG.Regulation of Trypanosoma cruzi infections in vitro and in vivo by transforming growth factor beta (TGF-beta). J Exp Med 1991; 174:539-545.
35. Freire-de-Lima CG, Nascimento DO, Soares, MBP, Bozza PT, Castro-Faria-Neto H, De Mello FG, DosReis GA, Lopes, MF. Uptake of apoptotic cells drives the growth of a pathogenic trypanosome in macrophages. Nature 2000; 403:1999-1203.
36. Munder M, Eichmann K, Moran JM, Centeno F, Soler G, Modolell M. Thl/Th2-regulated expression of arginase isoforms in murine macrophages and dendritic cells. J Immunol 1999; 163:3771-3777.
37. Saeftel M, Fleischer B, Hoerauf A. Stage-dependent role of nitric oxide in control of Trypanosoma cruzi infection. Infect Immun. 2001; 69:2252-2259.
38. Martins GA, Cardoso MGA, Aliberti JCS, Silva JS. Nitric oxide-induced apoptotic cell death in the acute phase of Trypanosoma cruzi infection in mice. Immunol. Letters 1998; 63:113-120.
39. Venturini G, Salvati L, Muolo M, Colasanti M, Gradoni L, Ascenzi P. Nitric oxide inhibits cruzipain, the major papain-like cysteine proteinase from Trypanosoma cruzi. Biochem. Biophys. Res. Commun. 2000; 270:437-441.
40. Cazzulo JJ, Stoka V, Turk V. Cruzipain, the major cysteine proteinase from the protozoan parasite Trypanosoma cruzi. Biol. Chem. 1997; 378:1-10.
41. Mottram JC, Brooks DR, and Coombs GH. Roles of cysteine proteinases of trypanosomes and Leishmania in host-parasite interactions. Curr. Opin. Microbiol. 1998; 1:455-460.
42. Denicola A, Rubbo H, Rodriguez D, Radi R. Peroxynitrite-mediated cytotoxicity to Trypanosoma cruzi. Arch. Biochem. Biophys. 1993; 304:279-286.
43. Thompson L, Gadella FR, Peluffo G, Vercesi AE, Radi R. Peroxynitrite affects Ca2+ transport in Trypanosoma cruzi. Mol. Biochem. Parasitol. 1999; 98:81-91.
44. Aliberti JC, Machado FS, Gazzinelli RT, Teixeira MM, Silva JS. Platelet-activating factor induces nitric oxide synthesis in Trypanosoma cruzi-infected macrophages and mediates resistance to parasite infection in mice. Infect Immun. 1999a; 67:2810-2814.
45. Aliberti JC, Machado FS, Souto JT, Campanelli AP, Teixeira MM, Gazzinelli RT et al. beta-Chemokines enhance parasite uptake and promote nitric oxide-dependent microbiostatic activity in murine inflammatory macrophages infected with Trypanosoma cruzi. Infect Immun 1999b; 67:4819-4826.
46. Murphy PM, Baggiolini M, Charo IF, Hebert CA, Horuk R, Matsushima K, Miller LH, Oppenheim JJ, Power CA. International union of pharmacology. XXII. Nomenclature for chemokine receptors. Pharmacol Rev. 2000; 52:145-176.
47. Villalta F, Zhang Y, Bibb KE, Kappes JC, Lima MF. The cysteine-cysteine family of chemokines RANTES, MIP-lalpha, and MIP-lbeta induce trypanocidal activity in human macrophages via nitric oxide. Infect Immun. 1998; 66:4690-4695.
48. Talvani A, Ribeiro CS, Aliberti JC, Michailowsky V, Santos PV, Murta SM, Romanha AJ, Almeida IC, Farber J, Lannes-Vieira J, Silva JS, Gazzinelli RT. Kinetics of cytokine gene expression in experimental chagasic cardiomyopathy: tissue parasitism and endogenous IFN-gamma as important determinants of chemokine mRNA expression during infection with Trypanosoma cruzi. Microbes Infect. 2000; 2:851-866.
49. Aliberti JC, Souto JT, Marino AP, Lannes-Vieira J, Teixeira MM, Farber J, Gazzinelli RT, Silva JS. Modulation of chemokine production and inflammatory responses in interferon-gamma- and tumor necrosis factor-R1-deficient mice during Trypanosoma cruzi infection. Am J Pathol. 2001; 158:1433-1440.
50. Krutzik SR, Sieling PA, Modlin RL. The role of Toll-like receptors in host defense against microbial infection. Curr Opin Immunol. 2001; 13:104-108.
51. Ropert C, Gazzinelli RT. Signaling of immune system cells by glycosylphosphatidylinositol (GPI) anchor and related structures derived from parasitic protozoa. Curr Opin Microbiol. 2000; 3:395-403.
52. Camargo MM, Almeida IC, Pereira ME, Ferguson MA, Travassos LR, Gazzinelli RT. Glycosylphosphatidylinositol-anchored mucin-like glycoproteins isolated from Trypanosoma cruzi trypomastigotes initiate the synthesis of proinflammatory cytokines by macrophages. J Immunol. 1997a; 158:5890-5901.

53. Camargo MM, Andrade AC, Almeida IC, Travassos LR, Gazzinelli RT. Glycoconjugates isolated from Trypanosoma cruzi but not from Leishmania species membranes trigger nitric oxide synthesis as well as microbicidal activity in IFN-gamma-primed macrophages. J Immunol. 1997b; 159:6131-6139.
54. Ropert C, Almeida IC, Closel M, Travassos LR, Ferguson MA, Cohen P, Gazzinelli RT. Requirement of mitogen-activated protein kinases and I kappa B phosphorylation for induction of proinflammatory cytokines synthesis by macrophages indicates functional similarity of receptors triggered by glycosylphosphatidylinositol anchors from parasitic protozoa and bacterial lipopolysaccharide. J Immunol. 2001; 166:3423-3431.
55. Almeida IC, Camargo MM, Procopio DO, Silva LS, Mehlert A, Travassos LR, Gazzinelli RT, Ferguson MA. Highly purified glycosylphosphatidylinositols from Trypanosoma cruzi are potent proinflammatory agents. EMBO J. 2000; 19:1476-1485.
56. Procopio DO, Teixeira MM, Camargo MM, Travassos LR, Ferguson MA, Almeida IC, Gazzinelli RT. Differential inhibitory mechanism of cyclic AMP on TNF-alpha and IL-12 synthesis by macrophages exposed to microbial stimuli. Br J Pharmacol 1999; 127:1195-1205.
57. Campos MA, Almeida IC, Takeuchi O, Akira S, Valente EP, Procopio DO, Travassos LR, Smith JA, Golenbock DT, Gazzinelli RT. Activation of toll-like receptor-2 by glycosylphosphatidylinositol anchors from a protozoan parasite. J Immunol. 2001; 167:416-423.
58. Hofffman RA, Langrehr JM, Billiar TR, Curran RD, Simmons RL. Alloantigen-induced activation of rat splenocytes is regulated by the oxidative metabolism of L-arginine. J. Immunol.1990;145: 2220-2226.
59. Candolfi E, Hunter CA, and Remington JS. Mitogen and antigen specific proliferation of T cells in murine toxoplasmosis is inhibited by reactive nitrogen intermediates. Infect. Immun. 1994; 62:1995-2001.
60. Candolfi E, Hunter CA, and Remington JS. Roles of gamma interferon and other cytokines in suppression of the spleen cell proliferative response to concanavaalin A and toxoplasma antigen during acute toxoplasmosis. Infect. Immun. 1995; 63:751-756.
61. Sternberg, J. and McGuigan F. Nitric oxide mediates suppression of T cell responses in murine Trypanosoma brucei infection. Eur. J. Immunol. 1992; 22:2741-2744.
62. Rockett KA, Awburn MM., Rockett EJ, Cowden, WB and Clark IA, Possible role of nitric oxide in malarial immunosuppression. Parasite Immunol. 1994; 16:243-246.
63. Gregory SH, Wing EJ, Hoffman RA, and Simmons RL. Reactive nitrogen intermediates suppress the primary immunologic response to Listeria. J. Immunol. 1993; 150:2901-2909.
64. Duhé RJ, Evans GA, Erwin RA, Kirken RA, Cox GW, Farrar WL. Nitric oxide and thiol redox regulation of Janus kinase activity. Proc Natl Acad Sci USA 1998; 95:126-131.
65. Leproivre M, Chenais B, Yapo A, Lemaire G, Thelander L, and Tenu JP. Alterations of ribonucleotide reductase activity following induction of the nitrite-generating pathway in adenocarcinoma cells. J. Biol. Chem. 1990; 265:14143-49.
66. Drapier JC and Hibbs JB, Differentiation of murine macrophages to express nonspecific cytotoxicity for tumor cells results in L-arginine-dependent inhibition of mitochondrial iron-sulfor enzymes in the macrophage effector cells. J. Immunol. 1988; 140:2829-2838.
67. Togashi H, Sasaki M, Frohman E, Taira E, Ratan RR, Dawson TM, Dawson VL. Neuronal (type I) nitric oxide synthase regulates nuclear factor kappaB activity and immunologic (type II) nitric oxide synthase expression. Proc Natl Acad Sci U S A 1997; 94:2676-2680.
68. Berendji D, Kolb-Bachofen V, Zipfel PF, Skerka C, Carlberg C, and Kroncke KD. Zinc finger transcription factors as molecular targets for nitric oxide-mediated immunosuppression: inhibition of IL-2 gene expression in murine lymphocytes. Mol Med. 1999; 11:721-730.
69. Kim YM, Bombeck CA, and Billiar TR. Nitric oxide as a bifunctional regulator of apoptosis. Circ. Res. 1999; 84:253-256.
70. Bogdan C. The function of nitric oxide in the immune system. In Heidelberg MB, (ed). Handbook of Experimental Pharmacology: Nitric oxide. Springer; 2000: 443-493.
71. Taylor-Robinson AW, Liew FY, Severn A, Xu D, McScorley SJ, Garside P, Padron J, Phillips RS. Regulation of the immune response by nitric oxide differentially produced by T helper type 1 and T helper type 2 cells. Eur. J Immunol. 1995; 25:3229-3234.

72. Ianaro A, O'Donnell CA, Di Rosa M, and Liew FY. A nitric oxide synthase inhibitor reduces inflammation, down-regulates inflammatory cytokines and enhances interleukin-10 production in carrageenin-induced oedema in mice. Immunology 1994; 82:370-375.
73. James SL, Cheever AW, Caspar P et al. Inducible nitric oxide synthase-deficient mice develop enhanced type 1 cytokine-associated cellular and humoral immune responses after vaccination with attenuated Schistosoma mansoni cercariae but display partially reduced resistance. Infect Immun 1998; 66:3510-3518.
74. Karupiah G, Chen JH, Mahalingam S et al. Rapid interferon gamma-dependent clearance of influenza A virus and protection from consolidating pneumonitis in nitric oxide synthase 2-deficient mice. J Exp Med. 1998a; 188:1541-1546.
75. Kolb H, Kolb-Bachofen V. Nitric oxide in autoimmune disease: cytotoxic or regulatory mediator? Immunol Today 1998; 19:556-561.
76. Rothe H, Hartman B, Geerlings P et al. Interleukin-12 gene-expression of macrophages is regulated by nitric oxide. Biochem Biophys Res Commun 1996; 224: 159-163.
77. Mattner F, Fischer S, Guckes S et al.The interleukin-12 subunit p40 specifically effects of the interleukin-12 heterodimer. Eur J Immunol 1993; 23:2203-2208.
78. Diefenbach A, Schindler H, Röllinghoff M et al. Requirement for type 2 NO-synthase for IL-12 responsiveness in innate immunity. Science 1999; 284:951-955.
79. Sicher SC, Vazquez MA, Lu CY. Inhibition of macrophage Ia expression by nitric oxide. J Immunol 1994; 153:1293-1300.
80. Ribeiro-dos-Santos R, Rossi MA, Laus JL et al. Anti-CD4 abrogates rejection and reestablishes long-term tolerance to syngenic newborn hearts grafted in mice chronically infected with Trypanposoma cruzi. J Exp Med 1992; 175:29-39.
81. Tarleton RL, Zhang L. Chagas Disease etiology: Autoimmunity or parasite persistence? Parasitol Today 1999; 15:94-99.
82. Bellotti G, Brocchi EA, Morais MVA et al. In vivo detection of Trypanosoma cruzi antigens in hearts of patients with chronic Chagas heart disease. Am Heart J 1996; 131:301-307.
83. Higuchi MD, Ries MM, Aiello VD et al. Association of an increase in CD8+ T cells with the presence of Trypanosoma cruzi antigens in chronic, human, chagasic myocarditis. Am J Trop Med Hyg 1997; 56:485-9.
84. Higuchi MD. Human Chronic chagasic cardiomyopathy: Participation of parasite antigens, subsets of lymphocytes, cytokines and microvascular abnormalities. Mem Inst Oswaldo Cruz 1999; 94(Suppl I):263-267.
85. Chandrasekar B, Melby PC, Troyer DA et al. Temporal expression of pro-inflammatory cytokines and inducible nitric oxide synthase in experimental acute chagasic cardiomyopathy. Amm J Pathol 1998; 152:925-934.
86. Ungureanu-Langrois D, Balligand JL, Kelly R et al. Myocardial contractile dysfunction in the systemic inflammatory syndrome: role of a cytokine-inducible nitric oxide synthase in cardiac myocytes. J Mol Cell Cardiol 1995; 27:155-167.
87. Huang H, Chan J, Wittner M et al. Expression of cardiac cytokines and inducible form of nitric oxide synthase (NOS2) in Trypanosoma cruzi-infected mice. J Mol Cell Cardiol 1999; 31:75-88.
88. Machado FS, Martins GA, Aliberti JC et al. Trypanosoma cruzi-infected cardiomyocytes produce chemokines and cytokines that trigger potent niytric oxide-dependent trypanocidal activity. Circulation 2000; 102:3003-3008.
89. Joe EK, Schussheim AE, Longrois D et al, Regulation of cardiac myocyte contractile function by inducible nitric oxide synthase (iNOS): mechanisms of contractile depression by nitric oxide. J Mol Cell Cardiol 1998; 30:303-315.
90. Ing DJ, Zang J, Dzau VJ et al. Modulation of cytokine-induced cardiac myocyte apoptosis by nitric oxide, Bak, and Bcl-x. Circ Res 1999; 84:21-33.
91. Pinsky DJ, Aji W, Szaboles M et al. Nitric oxide triggers programmed cell death (apoptosis) of adult rat ventricular myocytes in culture. Am J Physiol 1999; 277:H1189-199.
92. Rossi MA, Souza AC. Is apoptosis a mechanism of cell death of cardiomyocytes in chronic chagasic myocarditis? Inter J Cardiol 1999; 68:325-331.
93. Malachias LCC, Goldberg SS, Silva-Pereira AA et al. Role of Trypanosoma cruzi lipopolysaccharide on human granulocyte biologicalactivities. Mem Inst Oswaldo Cruz 1991; 86:469-470.

94. Aliberti JS, Cardoso MGA, Martins GA et al. IL-12 mediates resistance to Trypanosoma cruzi in mice and is produced by murine macrophages in response to live trypomastigotes. Infect Immun 1996; 64:1961-1967.
95. Rossi MA, Bestetti RB. The challenge of Chagas cardiomyopathy: the pathologic roles of autonomic abnormalities, autoimmune mechanisms and microvascular changes, and therapeutic implications. Cardology 1995; 86:1-7.
96. Bestetti BB, Rossi MA. A rationale approach for mortality risk stratification in Chagas heart disease. Int J Cardiol 1997; 58:199-209.
97. Balligand J, Ungureanu-langrois D, Simmons WW et al. Cytokine-inducible nitric oxide synthase (iNOS) expression in cardiac myocytes: Characterization and regulation of iNOS activity in single cardiac myocytes in vitro. J Biol Chem 1994; 269:27580-27588.
98. Ishiyama S, Hiroe M, Nishikawa T et al. Nitric oxide contributes to the progression of myocadial damage in experimental autoimmune myocarditis in rats. Circulation 1997; 95:489-496.
99. Petray PB, Castaños-Velez E, Grinsteins S et al. Role of nitric oxide in resistance and histopathology during experimental infection with Trypanosoma cruzi. Immunol Lett 1995; 47:121-126.
100. Kubes P, Suzuki M, Granger DN. Nitric oxide: an endogenous modulator of leukocyte adhesion. Proc Natl Acad Sci USA 1994; 88:4651-4655.
101. Adams MR, Jessup W, Hailstones D et al. L-arginine reduces human monocyte adhesion to vascular and endothelial expression of cell adhesion molecules. Circulation 1997; 95:553-4.
102. Whiss PA, Anderson RG, Srinivas U. Modulation of P-selectin expression on isolated human platelets by an NO donor assessed by a novel ELISA application. J Immunol Methods 1997; 200:135-143.
103. Astrup F, Vestweber D, Borges E et al. P- and E- selectin mediate recruitment of T-helper-1 but not T-helper-2 cells into inflammed tissues. Nature 1997; 385:81-83.
104. MacLean A, Wei XQ, Huang FP et al. Mice lacking inducible nitric oxide synthase are more susceptible to herpes simplex virus infection despite enhanced Th1 cell response. J Gen Virol 1998; 78:825-830.
105. Noda S, Tanaka K, Sawamura S et al. Role of nitric oxide synthase type 2 in acute infection with murine cytomegalovirus. J Immunol 2001; 166:3533-3541.
106. Karupiah G, Chen JH, Nathan CF et al. Identification of nitric oxide synthase 2 as an innate resistance locus against ectromelia virus infection. J Virol 1998b; 72:7703-7706.
107. Seydel KB, Smith SJ, Stanley SL Jr. Innate immunity to amebic liver abscess is dependent on gamma interferon and nitric oxide in a murine model of disease. Infect Immun 2001; 68:400-402.
108. McInnes IB, leung B, Wei X-Q et al. Septic arthritis following Staphylococcus aureus infection in mice lacking inducible nitric oxide synthase. J Immunol 1998; 160:308-315.
109. Rottenberg ME, Rothfuchs ACG, Gigliotti D et al. Role of imnnate an adaptative immunity in the outcome of primary infection with Chlamydia pneumoniae, as analyzed in genetically modified mice. J Immunol 1999; 162:2829-2836.
110. Tsai WC, Strieter RM, Zisman DA et al. Nitric oxide is required for effective innate immunity against Klebsiella pneumoniae. Infect Immun 1997; 65:1870-1875.
111. Millar AE, Sternberg J, McSharry C et al. T cell responses during Trypanosoma brucei infections in mice deficient in inducible nitric oxide synthase. Infect Immun 1999; 67:3334-3338.
112. Ramsey KH, Miranpuri GS, Poulson CE et al. Inducible nitric oxide synthase does not affect resolution of murine chlamydial genital tract infections or eradication of chlamydiae in primary murine cell culture. Infect Immun 1998; 66:835-838.
113. Perry LL, Feilzer K, Caldwell HD. Neither interlekin-6 nor inducible nitric oxide synthase is required for clearance of Chlamydia trachomatis from the murine genital tract epithelium. Infect Immun 1998; 66:1265-1269.
114. Myamoto Y, Akaike T, Kuwahara H et al. Urease function as a defense system of Helicobacter pylori against peroxynitrite through production of carbon monoxide [abstract]. Acta Physiol Scand 1999; 167(suppl 645):17.
115. Way SS, Goldberg MB. Clearance of Shigella flexneri infection occurs through a nitric oxide-independent mechanism. Infect Immun 1998; 66:3012-3016.
116. Hertz CJ, Mansfield JM. IFN-γ-dependent nitric oxide production is not linked to resistance in experimental african trypanosomiasis. Cell Immunol 1999; 192:24-32.

117. Favre N, Ryffel B, Rudin W. The development of murine cerebral malaria does not require nitric oxide production. Parasitology 1999; 118:135-138.
118. Yoneto T, Yoshimoto T, Wang C-R et al. Gamma interferon production is critical for protective immunity to infection with blood-stage Plasmodium berghei XAT but neither NO production nor NK cell activation is critical. Infect Immun 1999; 67:2349-2356.
119. Van der Heyde HC, Gu Y, Zhang Q et al. Nitric oxide is neither necessary nor sufficient for resolution of Plasmodium chabaudi malaria in mice. J Immunol 2000; 165:3317-3323.
120. Brown CR, Reiner SL. Development of lyme arthritis in mice deficient in inducible nitric oxide synthase. J Infect Dis 1999; 179:1573-1576.

Chapter 7

Impact of Polyclonal Lymphocyte Responses on Parasite Evasion and Persistence

Paola Minoprio

Introduction

In the field of immunoparasitology it is generally accepted that the survival and degree of pathogenicity of parasites is inextricably linked to their ability to escape and resist immune responses. Thus, numerous strategies have been proposed to explain the ability of micro-organisms to avoid immune surveillance and to persist within the host. However, the complex interactions between an infectious agent and its host that take place after infection and ultimately determine the state of susceptibility, or resistance to the development of pathology do not only rely on the "creativity" of micro-organisms. The genetic background of the host, the diversity, dynamics and activities of the immune system, and the events that precede 'immune reactions', are fundamental in determining the future of an infectious process.

In this Chapter I shall briefly review our own work in this area and present evidence for a relationship between the magnitude of polyclonal lymphocyte activation triggered by *Trypanosoma cruzi* parasites and the decreased resistance of the host to infection and to the development of tissue pathology. It is also apparent that polyclonally activated immune systems respond weakly to antigenic challenges as long as the state of activation is maintained. This may seem paradoxical. However this state of refractoriness/anergy is invariably and intimately associated with high levels of lymphocyte activation. This seems to be a consequence of the fact that very few of the multitude of lymphocyte clones that are generated in response to infection are directed to parasite epitopes. Thus the immune response that is induced is predominantly 'non-specific'. Interestingly, of the many functionally aberrant characteristics displayed by the immune system after infection with *T. cruzi*, polyclonal B- and T-cell non-specific responses and immunosuppression have not been sufficiently considered so far in the context of vaccination strategy and design. This Chapter is specifically aimed at addressing these questions and will attempt to outline new insights into the immunoresponses to parasites that have been provided by our recent findings.

Quantification of Total B- and T- Cell Responses after *Trypanosoma cruzi* Infection

In contrast to the majority of studies that aim to identify parasite specific B- and T-cell responses induced by *T. cruzi* infection, we have focussed our efforts on defining the global lymphocyte activity triggered by the parasite.[1] For this purpose it was important to quantify the changes that occur to the immune system during the course of the acute and chronic phases of murine infection regardless of clonal specificity. Using several imbred strains of mouse and

Molecular Mechanisms in the Pathogenesis of Chagas Disease, edited by John M. Kelly.

different strains of *T. cruzi,* we compared the immune responses to bloodstream forms of the parasite after intraperitoneal inoculation. Apart from some differences in the kinetics of lymphocyte responses, tissue inflammatory reactions and death, only very slight differences in major 'immunodisturbances' were detected. We observed the presence of large numbers of activated T- and B-lymphocytes very early after parasite injection, whose magnitude was incompatible with oligoclonal (parasite-directed) responses.[2] Infection of mice with the bloodstream forms of the parasite leads to a massive blast transformation of all lymphocyte classes both in the spleen and lymph nodes of 'resistant' or 'susceptible' mouse strains. The extent of blastogenesis was indicative that lymphocytes were activated by mechanisms independent of their recognition of antigenic parasite determinants and suggested that this activation could well result from polyclonal stimuli (see Fig. 1). Although the analysis of T- and B-cells responding to infection had already indicated the polyclonal non-specific nature of the cells that respond to *T. cruzi,*[3-9] the subsequent analyses of the whole T- and B-cell repertoires constituted the first molecular evidence for the polyclonality of those responses.[10,11] Thus, we demonstrated that B- and T-cells expressing all major V_H- and $V\beta$-gene families are stimulated and expanded during the first two weeks of parasite infection. Nevertheless, it should be noted that these cells mature and differentiate into effector cells that in the majority of cases produce antibodies or T-cell activities lacking parasite specificity.[5,9]

This massive polyclonal lymphocyte activation is the major immunological phenomenon observed after infection with *T. cruzi* and many other parasite, viral or bacterial pathogens.[12] More than half of the total cells of the immune system can be in a state of terminal differentiation. However, since these cells fail to bind to parasite antigens the response is essentially non-specific. There are obvious consequences associated with these observations. The first relates to the poor results that have been obtained with the approaches used to develop useful tools (or parasite epitopes) for immunotherapy or protective immunity. Since the major part of the response induced by the infection is devoid of 'parasite specificity' it is not surprising, in hindsight, that polyclonal activation of lymphocytes is a significant obstacle to the development of a good antigenic target for vaccination. As a result the rate of progress in this area of medical science has been disappointing, not only for *T. cruzi,* but for most other infections.

Immunosuppression: A Major Consequence of Polyclonal Activation

The mechanisms that give rise to what I have termed 'polyclonal non-specific responses' could alternatively result from specific activation which is merely representative of the sum of all specific responses to the multitude of pathogen antigens, thereby generating such high levels of activity.[13] A consequence of this would be that every single clone activated by the infection should react to parasite antigens. This hypothesis is therefore unlikely to be correct, since it should correlate with an outcome in which there is a better control of parasitemia in the initial phases of infection due to the performance of activated cells responding to the many parasite antigens. Consistent with this, we have shown that hypergammaglobulinemia follows the massive activation of B-cells after *T. cruzi* infection and that more than 95% of the secreted immunoglobulins fail to bind to parasite surface or intracellular antigens. By inference, most of the lymphocyte activity following infection is therefore largely independent of V-region specificity of the responding B-cell populations. This early lymphocyte hyperactivity is closely associated with a condition in which there is increased numbers of cells in secondary lymphoid organs—splenomegaly and lymphoadenopathy[14]—substantiating the profound commitment of the immune system machinery. Invariably, the engagement of most lymphocyte populations in effector activities that are not clonally specific leads to a state of impaired humoral and cellular responses to homologous and heterologous antigens.[15-21] This apparent reduced availability of clones able to respond to parasite and unrelated antigens may be readily explained by conventional immunosuppression.

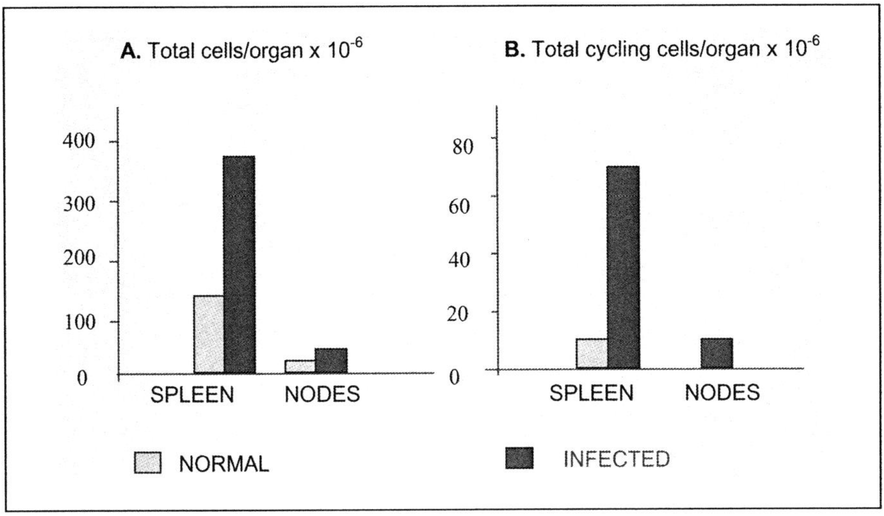

Figure 1. Massive lymphocyte proliferation and activation after *T. cruzi* infection. Lymphocyte population analysis in spleen and lymph nodes obtained from normal or 10 days infected C57BL/6 mice. Total number of lymphocytes/organ is depicted in panel A. Lymphoblastogenic reaction upon infection is shown in panel B where values represent total number of cycling lymphocytes (cells in S+G2+M phases of the mitotic cycle, as determined by cellular DNA content using a FACS).

Immunosuppression has always been considered as one of the major disorders induced by micro-organisms. Our hypothesis to explain this phenomenon is that the impaired capacity of lymphocytes to 'respond' to antigenic ligands after infectious processes is a reflection of the overwhelming activation of lymphoid cells triggered by polyclonal activators expressed by the micro-organisms. Polyclonally activated B- and T-lymphocytes enter a refractory state that disables the ability of these cells to respond to new stimuli. This association between polyclonal lymphocyte activation and immunosuppression certainly contributes to the reduction in host resistance and facilitates parasite evasion and persistence. In the case of *T. cruzi* infection, several experiments support the fact that increased resistance might be observed by reducing the levels of immune system engagement; decreasing B- and T- polyclonal responses has been shown to alleviate the resultant immunosuppression.[1,22-26] Although intriguing, the intense immune reaction that occurs early after infection may determine, at least partially, the outcome of infection. In addition it could allow the autonomous maintenance of hyperactivity of the immune system and thus contribute to the development of pathology. Besides, it is very likely that resistance to infection requires the differential participation of different classes of T-cell responses (Th1 type or Th2 type responses) that could regulate, or facilitate polyclonality (bystander effect of cytokines).[23,27,28] In *T. cruzi* infection, CD4, CD8 and double-negative T-cells bearing $\alpha\beta$ or $\gamma\delta$ T-cell receptors are all generally activated and an important fraction of these cells are implicated in non-specific cytotoxic, helper/regulator and suppressive functions.[1,2,29-32] Obviously if vaccination primes for the inappropriate class of response, rather than protect, it may facilitate infection and accelerate pathology (see **Box 1** for a summary).

Undesirable Polyclonal Cell Activation in Vaccination Approaches

Taken together all the above considerations suggest that a variety of cellular interactions take place after infection involving, directly or indirectly, different populations of B- and T-cells. This contributes to the predominant "non-specific" response to infection. It should also

Box 1. Principal immunogical dysfunction following* Trypanosoma cruzi *infection

- Polyclonal lymphocyte activation
  - All B- and T-lymphocyte classes (major and minor populations)
  - Hypergammaglobulinemia
  - Predominance of nonspecific B and T cell responses
- General Immunosuppression
  - Impaired response to homologous and heterologous antigens
- Progressive Autoimmunity
  - Expansion of self reactive B- and T- cell clones
  - Hyper-stimulation of the immune system and molecular mimicry

Question : If resistance to 'infection' and to 'pathology' requires different classes of T-cell responses (Th1 and Th2), what are the rules to be followed to trigger an appropriate class of T- cell response when choosing a vaccine candidate ?

be noted that specific responses directed to the parasite do exist, however the efficiency of these specific B- and T-cell activities are essentially submerged by the extent of polyclonal non-specific responses. It would seem therefore that *T. cruzi* employs, as do a variety of other micro-organisms, mitogenic and/or superantigenic products to elicit poor specific host responses as a mechanism to ensure survival.[33-47] Although some present vaccination protocols aim at inducing an immunological state of memory using parasite moieties that would simply lead to 'clinical immunity' (in spite of infection), it would seem difficult to envisage the development of 'effective' epitopic vaccines without neutralizing the deleterious effects of 'non-specific' polyclonal responses. Consequently, if 'sterile immunity' is not observed upon infection mitogens and/or superantigenic moieties released by the parasite would trigger polyclonal activation of B- and/or T-cells and systematically bring about immunosuppression. The consequences are relevant for vaccination purposes. On one hand, the failure to control B- and or T-cell hyperactivity could lead to the expansion of auto-reactive clones and a break-down of tolerance resulting in unwanted, progressive tissue damage. On the other hand, it could result in the abrogation of the beneficial effects of vaccination due to the immunosuppression that follows natural infection.

TcPA45: A Polyclonal B-Cell Mitogen Secreted by *Trypanosoma cruzi*

We have recently suggested that new strategies to generate protective immunity against *T. cruzi* should involve the isolation of moieties responsible for the triggering of non-specific polyclonal activation.[12] Previous studies had suggested that both mitogenic and superantigenic activities are associated with *T. cruzi* infection. The evidence that similar responses have been linked to infection with many other parasite, viral and bacterial pathogens has led to the proposal that these activities may represent a major strategy used by micro-organisms to avoid destruction by host defenses. Also of relevance are the observations that several immunopathological conditions have been associated with mitogenic/superantigenic–induced polyclonal activation, namely splenomegaly, adenopathy, toxic shock syndromes and progressive autoimmunity.

The impact of these observations on the susceptibility to *T. cruzi* infection encouraged us to use biochemical and molecular approaches to isolate the parasite products responsible for the initiation of non-specific polyclonal activation of host lymphocytes. We reasoned that

Table 1. Mitogenicity of TcPA45 in vivo

	Total Number of Ig-Producing B Cells		
	IgM-producing B cells	IgG2a-producing B cells	IgG2b-producing B cells
Non-injected	163000 ± 25456	3650 ± 636	4300 ± 142
TcPA45 (50 μg/mouse)	371666 ± 94495	9866 ± 3000	14633 ± 3287

	Total Number of Ig-Producing B Cells Anti-TcPA45		
	IgM-B cells anti TcPA45	IgG2a-B cells anti TcPA45	IgG2b-B cells anti TcPA45
Non-injected	None	None	None
TcPA45 (50 μg/mouse)	84	None	None

BALB/c mice were injected or not with 50 μg of recombinant TcPA45 (i.p.) and spleen cells assayed day 7 later. The results shown represent total numbers of spleen cells, total number of B cells producing IgM, IgG2a or IgG2b isotypes, and total numbers of isotype-producing B cells specific to the protein. The crucial observations are that there is 1) a 2 fold increase in total spleen cells numbers 7 days after injection with rTcPA45, 2) a 3-6 fold increase in total numbers of Ig-secreting cells and 3) less than 0.5% of IgM secreting cells are directed to the injected protein, characterizing a mitogenic stimulation of B cells.

polyclonal activation of B-cells by parasite mitogens could contribute to the class determination of T-cell responses and to the ensuing T-cell suppression and anergy observed after infection. Using culture supernatants of metacyclic forms of the parasite (similar to the infectious stage transmitted by the insect vector), we hypothesized that a possible strategy for *T. cruzi* to evade the host immune system could be reliance on secretion of mitogenic substances at the moment of the infection. This was indeed found to be the case. Following a series of rigorous experiments we were able to purify and identify a secreted parasite protein of 45 kDa involved in polyclonal B-cell activation.[48] TcPA45 (for *Trypanosoma cruzi* Polyclonal Activator) was shown to be a T-cell independent B-cell mitogen both in vitro and in vivo (Table 1) (see Box 2 for complementary information on B cell biology).

Interestingly, we found that the TcPA45 mitogen is a proline racemase, and were able to demonstrate that it could catalyze the interconversion of L- and D-enantiomers of proline (Fig. 2). To our knowledge this was the first report of an eukaryotic proline racemase gene. Of major significance in a chemotherapeutic context, our results indicated that the integrity of the enzyme active site is necessary for B-cell mitogenic activity. Thus, mitogenic activity is abolished, or severely compromised whenever enzymatic activity is inhibited by specific inhibitors or substrate (proline) excess. Although these data are suggestive that TcPA45 is directly mitogenic for B-cells, we cannot exclude the possibility that it can act through racemization of host molecules which could then function as the primary mitogens for B-cells. In this case it would be implicit that the activity of the TcPA45 racemase is dependent on peptide context. Regardless of the precise mechanism, non-specific activation of B-cells would be induced and provide the conditions which enable the parasite to evade immune destruction.

Box 2. Some words on the biology of B lymphocytes

- 90% of all B lymphocytes in naive individuals are "resting", non-cycling cells (G0 state of the mitotic cycle, or arrested in G1 phase)
- 5-10% of all B cells are large, activated B cells (blasts), even if rescued from germ-free or antigen-free individuals (cycling cells at S+G2+M phases of the mitotic cycle)
- In the resting state B lymphocytes do not express receptors for growth factors and are refractory to their reactivity
- The process of B-cell activation requires two different steps: a) induction, with acquisition of reactivity to (polyclonal) growth factors and b) proliferation and terminal differentiation to Ig-secreting cells (driven by response to growth and maturation factors)
- Polyclonal B-cell mitogens induce 5-50% of all B cells to proliferate and to secrete antibodies in the absence of any factor or accessory cell
- Each dividing B-cell gives rise to a clone containing up to 1000 progeny cells
- Upon stimulation, B-cells increase secretion of immunoglobulin by 100-1000 fold
- Terminal maturation and ultimate antibody secretion is incompatible with further growth (DNA synthesis can be blocked without decreases in Ig production)
- Upon mitogenic stimulation B lymphocyte growth arrest precedes maturation to high rate immunoglobulin-secretory cells

The finding that mitogenic activity is dependent on the availability of the racemase active site is indeed attractive as a possible new approach to therapeutic protocols. To date, only two drugs are available to treat acute *T. cruzi* infection. However, Chagas disease cannot be cured in its chronic stages or in its congenital form. As mitogenicity of TcPA45 is completely compromised by its previous incubation with pyrrole carboxylic acid, a planar active site specific inhibitor of proline racemase, the determination of the protein structure of proline racemase may allow the design of potential inhibitors to be used in human therapy. Given the fact that chronic tissue pathology has been attributed to the expansion of autoreactive B- and T-cell clones developed during polyclonal lymphocyte activation triggered in early phases of infection, proline-racemase active site inhibition would be one candidate target for drug design. Current studies are now in progress to verify if the proline racemase of *T. cruzi* also plays a role in the biological and metabolic processes of the parasite, and whether it is involved in parasite adhesion to host cells and in its consequent internalization and multiplication. The answer to these questions may provide additional information in the search for compounds to target the parasite proline racemase.

Parasite Evasion and Persistence Can Be Explained by Polyclonal Reactions

These findings have considerable implications for *T. cruzi* biology and pathogenicity, as well as for the role of D-amino acids in immune phenomena. Curiously, it has been shown that polypeptides composed exclusively of D-amino acids induce "immunological paralysis", persist for long periods and are resistant to enzymatic degradation.[49,50] The induction of specific B-cell responses to polypeptides composed of D-amino acids is inversely correlated to the injected dose. Accordingly, it was demonstrated that multi-chain polypeptides composed of D-amino acids induce antibody responses in a T-cell independent manner, equivalent to the response following B-cell responses triggered by mitogens.[49] It is well know that mitogens are very "immunogenic" but do not act as very good "antigens".[49,51] This was demonstrated by the analysis of antibody responses induced after mitogenic injection (see Box 3). The resulting

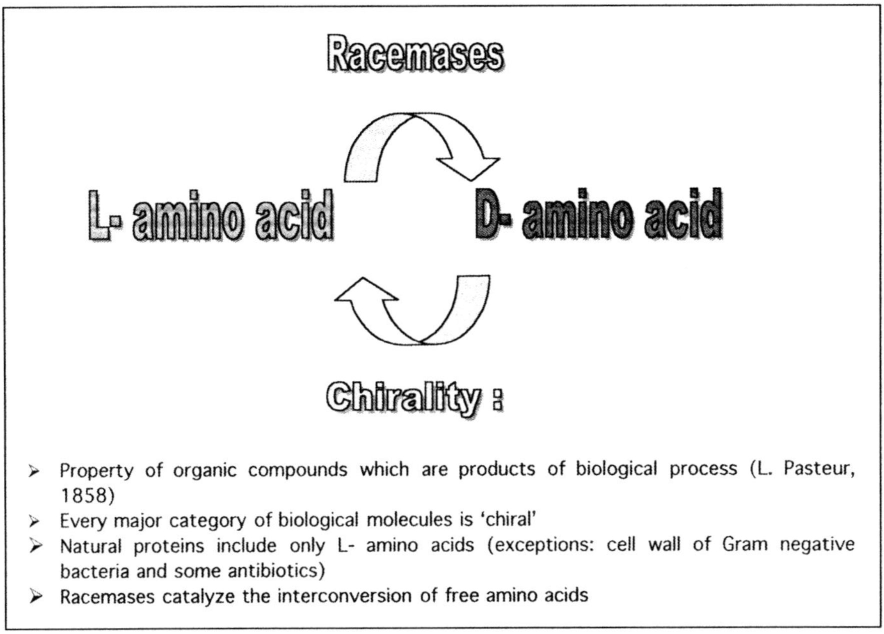

Figure 2. Racemase enzymes play important role in biological processes: synthesis of D-amino acids and conversion of nonmetabolizable to metabolizable isomer of an amino acid.

antibodies recognized very few, or none, of the inducer molecules. However, triggering of specific responses by mitogens is possible but, as with polypeptides composed of D-amino acids, it is inversely correlated to the injected dose. Injection of low doses of mitogens is able to induce specific B-cell responses to determinants in the mitogenic molecule, while at high doses, the response is polyclonal and non-specific, and is followed by selective paralysis of specific B-cells. Thus, our recent results suggest that injection of sub-mitogenic doses of *T. cruzi* proline racemase is indeed able to induce specific antibody responses against the TcPA45 molecule and to confer significant resistance to subsequent lethal challenges with the parasite (see ref. 48 and manuscript in preparation).

It would seem therefore that the generation of specific responses to these secreted mitogens may be responsible for the neutralization of parasite induced polyclonal activation at the very beginning of an infection. This would appear to overcome the parasite survival strategy based on the non-specific proliferation of B-cells. It is tempting to speculate that systematic experimental analyses using mitogens as 'immunogens' might reveal that the induction of effective immune protection can be attained by using 'mitogenic moieties' in association with more conventional candidate epitopes. If correct, this hypothesis predicts that the triggering of specific responses to essential parasite molecules, in the absence of non-specific polyclonal responses, should be successful in preventing infection (sterile immunity). It further suggests that this approach could prevent the indirect reactions involved in the progress of chronic pathology.

Concluding Remarks

Experimental Chagas disease provides an excellent model to investigate the initial phases of infection and the development of an immune response. The model also allows in vivo studies of the immune system in its totality permitting basic questions to be addressed, such as the inadequacies of the response that follow most infections with micro-organisms, including

Box 3. Quantification of mitogenic B-cell responses

As B-cells are activated by mitogens regardless of their immunological specificity, some principles to analyze and quantify mitogenic responses should include :

- Determination of the number of B-cells initiating growth upon mitogenic stimulation (i.e. frequency of total reactive B-cells)
- Determination of the number of times each responding cell clone divides (extent of proliferation)
- Determination of the efficiency of terminal differentiation (number of resulting Ig-secreting cells)
- Determination of the quality of the response : specificity and amount of Ig produced (ratio specific mitogen-directed Ig / total secreted Ig)
- Analysis of B-cell repertoire gene families : indiscriminate utilization of V_H-genes following B-cell activation is incompatible with 'specific', oligoclonal responses

immunosuppression and pathological autoimmunity. The inappropriate responses following infection with *T. cruzi* can be triggered if 'sterile immunity' is not achieved. Of more concern is the possibility that trials to induce 'immunity' against an infection by vaccination using 'conventional' immunodominant parasite molecules, although leading to some degree of protection, could also be responsible for triggering mechanisms implicated in the development of the late chronic disease. Consequently 'vaccination' could be worse than no protection at all, since it could accelerate the onset and exacerbate the severity of disease pathology. In addition autoimmune reactivities induced by the 'vaccine' could be boosted by 'natural' challenge. We have proposed that the difficulties faced by the immune system in eliminating micro-organisms could stem from the fact that micro-organisms trigger non-specific polyclonal B- and T- lymphocytes that interfere with the raising of an efficient immune response early in infection. Classical vaccination approaches have focused on the study of "immunodominant" or "immunopathological" epitopes, although it is now clear that the great majority of the immunologically relevant interactions that occur after infection are not specific. We therefore suggest that better approaches to inducing immunity should involve the isolation of the moieties responsible for triggering non-specific polyclonal lymphocyte responses, the neutralization of their activities in the host, or their genetic inactivation in the micro-organism. In so doing we may improve our knowledge of the interactions of the host immune system with the parasite and reveal general principles involved in the development of progressive disease following infection.

References

1. Minoprio P, Itohara S, Heusser C et al. Immunobiology of murine T. cruzi infection: The predominance of parasite-nonspecific responses and the activation of TcRI T cells. Immunol Rev 1989; 112:183-207.
2. Minoprio P, Eisen H, Forni L et al. Polyclonal lymphocyte responses to murine Trypanosoma cruzi infection. I. Quantitation of both T and B cell responses. Scand J Immunol 1986; 24:661-668.
3. d'Imperio-Lima MR, Joskowicz M, Coutinho A et al. Very large and isotypically atypical polyclonal plaque-forming cell responses in mice infected with Trypanosoma cruzi. Eur J Immunol 1985; 15:201-203.
4. d'Imperio-Lima MR, Eisen H, Minoprio P et al. Persistance of polyclonal B cell activation with undetectable parasitemia in late stages of experimental Chagas disease. J Immunol 1986; 137:353-356.

5. Minoprio P, Burlen O, Pereira P et al. Most B cells in acute Trypanosoma cruzi infection lack parasite specificity. Scand J Immunol 1988; 28:553-561.
6. Spinella S, Liegeard P, Hontebeyrie-Joskowicz M. Trypanosoma cruzi: Predominance of IgG2a in nonspecific humoral response during experimental Chagas disease. Exp Parasitol 1992; 74:46-56.
7. Spinella S, Milon G, Hontebeyrie-Joskowicz M. A CD4+ TH2 cell line isolated from mice chronically infected with Trypanosoma cruzi induces IgG2 polyclonal response in vivo. Eur J Immunol 1990; 20:1045-1051.
8. Ortiz-Ortiz L, Parks DE, Rodriguez M et al. Polyclonal B lymphocyte activation during Trypanosoma cruzi infection. J Immunol 1980; 124:121-126.
9. Minoprio P, Coutinho A, Joskowicz M et al. Polyclonal lymphocyte responses to murine Trypanosoma cruzi infection. II. Cytotoxic T lymphocytes. Scand J Immunol 1986; 24:669-679.
10. Leite-de-Moraes MdC, Coutinho A, Hontebeyrie-Joskowicz M et al. Skewed Vb TCR repertoire of CD8+ T cells in murine Trypanosoma cruzi infection. Int Immunol 1994; 6:387-392.
11. Minoprio P, Andrade L, Lembezat M-P et al. Indiscriminate representation of V_H-gene families in the murine B lymphocyte responses to Trypanosoma cruzi. J Immunol 1989; 142:4017-4021.
12. Reina-San-Martin B, Cosson A, Minoprio P. Lymphocyte polyclonal activation: A pitfall for vaccine design against infectious agents. Parasitol Today 2000; 16:62-67.
13. Arala-Chaves M, Lima MR, Coutinho A et al. V-region-related and -unrelated immunosuppression accompanying infections. Mem Inst Oswaldo Cruz 1992; 87:35-41.
14. Andrade ZA. Pathogenesis of Chagas disease. Res Immunol 1991; 142:126-129.
15. Hansen DS, Alievi G, Segura E et al. The flagellar fraction of Trypanosoma cruzi depleted of an immunosuppressive antigen enhances protection to infection and elicits spontaneous T cell responses. Parasite Immunol 1996; 18:607-615.
16. Kierszenbaum F. What are T-cels subpopulations really doing in Chagas disease? Parasitol Today 1995; 11:6-7.
17. Kierszenbaum F, Lopez HM, Sztein MB. Inhibition of Trypanosoma cruzi-specific immune responses by a protein produced by T. cruzi in the course of Chagas disease. Immunol 1994; 81:462-467.
18. Plata F. Enhancement of tumor growth correlates with suppression of the tumor- specific cytolytic T lymphocyte response in mice chronically infected by Trypanosoma cruzi. J Immunol 1985; 134:1312-1319.
19. Ramos C, Lamoy E, Feoli M et al. Trypanosoma cruzi: Immunosupressed response to different antigens in the infected mouse. Exp Parasitol 1978; 4:190-199.
20. Tarleton RL, Kuhn RE. Measurement of parasite-specific immune responses "in vitro": Evidence for suppression of the antibody response to Trypanosoma cruzi. Eur J Immunol 1985; 15:845-850.
21. Harel-Bellan A, Joskowicz M, Fradelizi D et al. Modification of T-cell proliferation and interleukin 2 production in mice infected with T. cruzi. Proc Natl Acad Sci USA 1983; 80:3466-3469.
22. Minoprio P, Eisen H, Joskowicz M et al. Suppression of polyclonal antibody production in Trypanosoma cruzi infected mice by treatment with anti-L3T4 antibodies. J Immunol 1987; 139:545-550.
23. Minoprio P. Chagas disease: CD5 B-cell dependent Th2 pathology? Res Immunol 1991; 142:137-140.
24. Minoprio P, Cury-El-Cheikh M, Murphy E et al. Xid-associated resistance to experimental Chagas disease is IFN-g-dependent. J Immunol 1993; 151:4200-4208.
25. Santos Lima EC, Vasconcelos R, Reina San Martin B et al. Significant association between the skewed natural antibody repertoire of xid mice and resistance to Trypanosoma cruzi infection. Eur J Immunol 2000; 31:634-645.
26. Santos-Lima EC, Minoprio P. Chagas disease is attenuated in mice lacking gd T cells. Infec Immun 1996; 64:215-221.
27. Cury-El-Cheikh M, Hontebeyrie-Joskowicz M, Coutinho A et al. CD5 B cells: Potential role in the (auto)immune responses to Trypanosoma cruzi infection. In: Herzenberg L, Haughton G, herzenberg L, Rajewisky K, eds. CD5 B cells in Development and Disease. New York: The New York Academy of Sciences, 1992:557-563.
28. Cordeiro-da-Silva A, Lima ECS, Vicentelli M-H et al. Vb6-bearing cells are involved in resistance to Trypanosoma cruzi infection in XID mice. Internat Immunol 1996; 8:1213-1219.

29. Minoprio P, Bandeira A, Pereira P et al. Preferential expansion of Ly1-B and CD4⁻ CD8⁻ T cells in the polyclonal lymphocyte responses to murine Trypanosoma cruzi infection. Intern Immunol 1989; 1:176-184.
30. Reed S, Brownell CE, Russo DM et al. IL-10 mediates susceptibility to Trypanosoma cruzi infection. J Immunol 1994; 153:3135-3140.
31. Santos-Lima EC, Garcia I, Vicentelli M-H et al. Evidence for a protective role of tumor necrosis factor in the acute phase of Trypanosoma cruzi infection in mice. Infect Immun 1997; 65:457-465.
32. Cardillo F, Voltarelli JC, Reed SG et al. Regulation of Trypanossoma cruzi infection in mice by gamma interferon and interleukin 10: Role of NK cells. Infec Immun 1996; 64:128-134.
33. Anderson R. AIDS: Trends, predictions, controversy. Nature 1993; 363:393-394.
34. Arala-Chaves M, Ribeiro A, Santarem M et al. Strong mitogenic effect for murine B lymphocytes of an immunosuppressor substance released by Streptococcus intermedius. Infec Immunity 1986; 54:543-548.
35. Ardavin C, Luthi L, Andersson M et al. Retrovirus-induced target cell activation in the early phases of infection: The mouse mammary tumor virus model. J Virol 1997; 71:7295-7299.
36. Assoku RK, Tizard IR. Mitogenicity of autolysates of Trypanosoma congolense. Experientia 1978; 34(1):127-129.
37. Azuma M, Ikeda M, Noda K et al. Mitogenic activity of soluble preparations from Plasmodium berguei-infected erythrocytes to L3T4+, Lyt-2- and L3T4- Lyt3+ T-cell subsets. Zentralblatt Bakteriol 1990; 273:401-411.
38. Bakhiet M, Olsson T, Edlung C et al. A Trypanosoma brucei brucei derived factor that triggers CD8+ lymphocytes to interferon-γ secretion: Purification, characterization and protective effects in vivo by treatment with a monoclonal antibody against the factor. Scand J Immunol 1993; 37:165-178.
39. Bona C, Broder S, Dimitriu A et al. Polyclonal activation of human B lymphocytes by Nocardia water soluble mitogen (NWSM). Immunol Rev 1979; 45:69-92.
40. Burckhardt J, Guggenhein B, Hefti A. Are actinomyces viscosus antigens B cell mitogens? J Immunol 1977; 118:1460-1465.
41. Coutinho A, Moller G. B cell mitogenic properties of thymus-independent antigens. Nature—New Biol 1973; 245(140):12-14.
42. Diamantstein T, Trissl D, Klos M et al. Mitogenicity of Entamoeba histolytica extracts for murine lymphocytes. Immunol 1980; 41:347-352.
43. Fleisher B. Superantigens produced by infectious pathogens: Molecular mechanism of action and biological significance. Int J Clin Lab Res 1994; 24:193-197.
44. Herman A, Kappler JW, Marrack P et al. Superantigens: Mechanism of T-cell stimulation and role in immune responses. Ann Rev Immunol 1991; 9:745-772.
45. Johnson RB, Kohl S, Bessler WG. Polyclonal activation of B-lymphocytes in vivo by Salmonella typhimurium lipoprotein. Infec Immun 1983; 39:1481-1484.
46. Lohoff M, Matzner C, Röllinghoff M. Polyclonal B-cell stimulation by L3T4⁺ T cells in experimental Leishmaniasis. Infec Immunity 1988; 56:2120-2124.
47. Scherer MT, Ignatowicz L, Winslow GM. Superantigens: Bacterial and viral proteins that manipulate the immune system. Annu Rev Cell Biology 1993; 9:101-128.
48. Reina-San-Martin B, Degrave W, Rougeot C et al. A B-cell mitogen from a pathogenic trypanosome is a eukaryotic proline racemase. Nature Medicine 2000; 6:890-897.
49. Sela M, Zisman E. Different roles of D-amino acids in immune phenomena. FASEB J 1997; 11:449-456.
50. Janeway CA, Humphrey JH. Synthetic antigens composed exclusively of L- or D-amino acids. II. Effect of optical configuration on the metabolism and fate of synthetic polypeptide antigens in mice. Folia Biol 1970; 16:156-172.
51. Coutinho A, Gronowicz E, Bullock WW et al. Mechanism of thymus-independent immunocyte triggering. Mitogenic activation of B cells results in specific immune responses. J Exp Med 1974; 139:74-92.

CHAPTER 8

Activation of Bradykinin-Receptors by *Trypanosoma cruzi*:

A Role for Cruzipain in Microvascular Pathology

Julio Scharfstein

Abstract

During its life cycle in the mammalian host, *Trypanosoma cruzi* productively exploits the enzymatic diversity of its own proteases to generate activation signals for a broad range of host cells. At least for the host responses relayed by the G-protein coupled receptors, the generation of the signalling agonists depends on the processing of different inactive precursor molecules by unique parasite proteases. For example, the major cysteine proteinase of *T. cruzi,* cruzipain, is able to process kininogen molecules that are docked to heparan sulphate proteoglycans, thereby promoting release of vasoactive "kinin" peptides. Whether transduced by constitutive (B2) or inducible (B1) kinin-receptor subtypes, the vigorous $[Ca^{2+}]_i$ transients triggered by the short-lived kinin peptides drastically increase host cell susceptibility to trypomastigote invasion. Given the evidence that kinins are rapidly metabolized by host peptidases (e.g., kininase I and II), differences in the tissue levels of these kinin degrading enzymes may influence the extent of organ involvement and pathological outcome in individual patients. Furthermore, recent studies indicate that the kinin-activating phenotype is not ubiquitously expressed in the genetically diverse *T. cruzi* species. Analysis of the mechanisms underlying the kinin releasing activity of parasite strains/clones have tentatively linked this competence to the expression of particular subsets of cruzipain isoforms. Insight on the regulatory checkpoints involved in the activation of the vascular endothelium by *T. cruzi* may shed light on the pathogenesis of Chagas disease.

Introduction

The dissection of the signal transduction pathways by which intracellular protozoa modulate host cell responses after attachment to the cell surface has become a central theme in molecular parasitology. Insight into the molecular strategies which *T. cruzi* employs to survive in different hosts has been obtained from relatively simple analysis of parasite-host cell interactions in culture systems. The importance of cellular signalling to the parasite invasion process is particularly well documented. Depending on the nature of the infective forms, either metacyclics[1] or tissue-culture derived trypomastigotes[2] and of the parasite strain-host cell combination,[3] different activation pathways promote cellular invasion.[4-6b] Studies with tissue culture trypomastigotes (TCT) have pointed to the involvement of the TGF-β signal transduction

Molecular Mechanisms in the Pathogenesis of Chagas Disease, edited by John Kelly.

pathway,[5] but in other cell systems it was shown that parasite-derived signalling molecules trigger $[Ca^{2+}]_i$ influx by activating G-protein coupled receptor (GPCRs).[6] Moreno et al[7] have also observed that upon parasite attachment to host cells, there is a transient increase in the cytosolic $[Ca^{2+}]_i$ of trypomastigotes. These studies indicated that cellular communication is bi-directional at this early stage of the host-parasite interaction process. In an interesting twist, Pereira and co-workers[4] pointed out that host cell susceptibility (e.g., cardiac muscle, smooth muscle and skeletal muscle) to infection correlates inversely with parasite competence to activate NF-κB. These findings suggested that activation of this nuclear transcriptional factor by the pathogen may inhibit progression of intracellular parasitism at later times and conversely, that host cell types in which NF-κB is not activated upon parasite contact are intrinsically more susceptible to infection. Considering the extent of genetic and biological diversity which exists in natural isolates of *T. cruzi* (Miles, M.A., this publication), tissue tropism in the acute infection may influenced by the balance of activating signals relayed by multiply attached parasites.

Efforts to delineate the mechanisms by which trypomastigotes generate signalling molecules that stimulate $[Ca^{2+}]_i$ influx via activation of G protein-coupled receptors (GPCRs) have recently focussed on different classes of parasite proteases. For example, Andrews and co-workers presented several lines of evidence indicating that signalling molecules produced by TCT derived from proteolytic processing of an inactive polypeptide precursor by an oligopeptidase B-type of enzyme.[8] Since this peptidase is found in the cytoplasm of the parasites, it has been suggested that the newly formed agonists must be transported to extracellular spaces. A different scenario emerged when we examined the role of cruzipain, a lysosomal-like proteinase in host cell invasion mechanisms.[11,28,113] Given indications that cruzipain was able to release vasoactive "kinin" peptides from the kininogens,[10] we asked if trypomastigotes could mobilize this proteinase to generate activation signals for the kinin receptors.[11,113] These studies demonstrated that host cell susceptibility to parasite invasion was drastically increased due to the vigorous $[Ca^{2+}]_i$ responses which the released kinins induced through kinin-receptors.[11,113] In the following section, I will review current information on the structure and function of cruzipain, emphasizing the features that are relevant for pathogenesis research.

Structural and Enzymatic Properties of the Major Cruzipain Isoform

Assigned to the clade A of the C1 cysteine peptidase branch,[12] the papain-like cysteine proteinases from the cruzipain family are encoded by a large number of gene copies (up to 130 copies in the Tul 2 strain) arranged as tandem arrays on several chromosomes.[13] Similar to the closely related (type–1) proteases from *Trypanosoma brucei* and *Leishmania mexicana,*[14] members of the cruzipain family share a long C-terminal extension (CTE)[15] of unknown function, which is absent in lysosomal cathepsins from mammalian cells.

The examination of full-length cruzipain genes[13,16,17,17b] revealed that the corresponding proteins are synthesized as zymogens (Fig. 1) that undergo post-translational modifications[18] and autocatalytic processing[16] prior to delivery into lysosomes.[19,20] Analysis of the X-ray-derived structure of cruzain,[21] a truncated recombinant form of the major lysosomal protease (lacking the C-terminal domain) bound to an irreversible pseudopeptide inhibitor, has yielded clues about its substrate specificity. As reviewed by McGrath,[22] the overall folding pattern of the 215-amino-acid polypeptide chain of cruzain is similar to that of papain, the catalytic triad (residues Cys-25, His-162 and Asn-182) being likewise confined to an extended substrate-binding site located between two internal domains. Apart from shifts in the position of Cys-153/Cys-200 disulphide bonds and a buried Cys-36, the structural differences in relation to papain and mammalian cathepsin L/S are, as expected, more accentuated in regions of loops and turns. The S1 subsite pocket of cruzipain contributes to some extent to the determi-

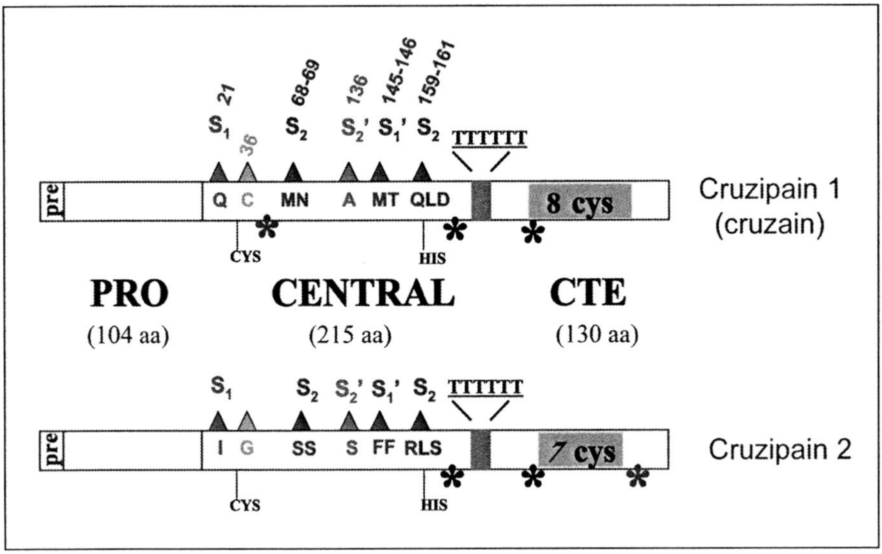

Figure 1. Schematic representation of cruzipain isoenzymes. The domain organization of preprocruzipain is illustrated. The active site residues, Cys-25 and His-162 of cruzain[13,16] (Genebank accession no. M84342 or M69121) and cruzipain 2 (accession no. M90067)[17] are marked in black. Subsite assignment was based on the X-ray structure of cruzain.[21] The catalytic domain (215 residues) is separated from the 130 amino acid long carboxy-terminal extension (CTE) by a proline-rich "hinge" sequence containing 7 threonine residues[15] (marked in gray). Non-conserved amino acid substitutions found in the S_1 subsite (21-23) are marked in pink while those present in the S_2 subsite (68, 69, 159-161) are marked in blue. Non conserved residues found in the vicinity of the S_1' sub-site (145 and 146) are marked in green. Those in the vicinity of the S2' subsite (136) is marked in red. Cys-36 (orange), which is unpaired in cruzain is not present in cruzipain 2 or in other members of the C1 peptidase family.[22]Both isoforms have 3 putative sites for N-glycosylation (*). The N-linked oligosaccharide chain[55] present in the N-terminal end of cruzain (Asn-47), hereafter termed cruzipain 1, is not present in cruzipain 2.[17] Two other N-linked glycosylation sites (Asn-170 and Asn-255) are shared by both isoforms. The cruzipain 2 sequence predicts a unique oligosaccharide N-linked site (Asn-321) at its C-terminal end. The 76 amino acid long internal core (green) of cruzipain 1 is stabilized by 4 pairs of disulfide bridges.[54] The last cysteine residue (Cys-318) of cruzipain 1 is missing in the CTE of cruzipain 2,[17] suggesting that its internal core may be shorter and the hydrophylic tail more exposed. Most of the antibody specificities present in chagasic sera are directed against the CTE of cruzipain,[36,37,41] originally defined as GP25, a diagnostic antigen.[36-38] T cell epitopes recognized by $CD4^+$ lymphocytes isolated from chronic chagasic patients were mapped to an internal peptide sequence in the catalytic domain.[45]

nation of its substrate specificity,[22] but similar to other papain-like enzymes, these interactions are less critical than those involving the S2 subsite. Although the overall structure of cruzipain is more closely related to cathepsin L, the parasite proteinase has a dual substrate specificity.[22,25] Lima et al[24] were the first to predict that Arg residues placed in the P2 position of the polypeptide substrates interact with Glu-205 (papain numbering) in the hydrophobic S2 specificity pocket, and hence could account for the cathepsin B-like specificity of cruzipain.[24] Comparative analysis of X-ray structures derived from different enzyme-inhibitor complexes[25] revealed that replacement of the positively charged P2 residue with a hydrophobic residue promotes a pH-dependent structural rearrangement of the S2 pocket residue Glu-205. This structural feature explains why cruzipain shares enzymatic properties with both cathepsin L and cathepsin B of mammalian origin.[9,24,26-30]

In spite of the wealth of information derived from structure-activity studies on cruzipain, our knowledge of cleavage sites from natural protein substrates is still limited to a few examples. The analysis of the auto-catalytic processing of pro-cruzipain has revealed homologies between two internal cleavage sites: the VVG-APA sequence which links the catalytic domain to the inhibitory pro-segment,[16] and VVG-GPG, the protease sensitive site that links the central domain to the CTE.[31] The contribution of the Pro residue as a determinant of P2' specificity, inferred from the above studies, was supported by multicombinatorial library screening analysis of the enzymes' substrate specificity.[29]

The possibility that kininogen could serve as a natural substrate for cruzipain was first suggested by Juliano and co-workers.[10] Kinetic assays performed with intramolecularly quenched substrates spanning the N-terminal and C-terminal flanking sequences from the bradykinin moiety of kininogens indicated that cruzipain and tissue kallikrein had somewhat overlapping specificities. Consistent with this, we have further observed that natural cruzipain liberates lysyl-bradykinin ("kallidin") from purified kininogens.[10] At first sight, this result was paradoxical because kininogens are members of the cystatin family of cysteine protease inhibitors, hence are able to reversibly inactivate papain-like enzymes such as cruzipain, in vitro.[32] As discussed later in this text, Lima et al[10b] have recently reported that interactions of heparan sulphate with high-molecular weight kininogen and cruzipain impairs the inhibitory activity of the cystatin-like domains.[10b] These cooperative interactions may facilitate the kinin-releasing activity of cruzipain in the in vivo settings.[10b]

Antigenic Properties of Cruzipain

For over a decade, research on seemingly unrelated *T. cruzi* antigens and proteinases evolved in parallel before they converged, in the early 1990s. After initial description of the proteases present in cell-free extracts from epimastigotes,[33] a cysteine proteinase of 60 kDa was identified by Rangel et al.[34] Cazzulo and co-workers were the first to characterize the major cysteine proteinase of epimastigotes as a cathepsin L-like lysosomal enzyme,[26,27] which they termed as "cruzipain".[27] While these enzymatic studies were in progress, Mendonça-Previato et al reported the isolation of GP25, a glycoprotein with mucin-like properties (40% w/w carbohydrate), from boiled extracts of epimastigotes.[35] We subsequently reported that chagasic patients mount prominent antibody responses to GP25 antigen.[36,37] Immunochemical analysis of ^{35}S-metabolically labelled proteins later identified epimastigote glycoproteins of higher molecular mass (GP57/51) as the primary biosynthetic products from which the 25-kDa fragments (i.e., GP25) were derived by proteolysis.[38] N-terminal sequencing, enzyme kinetic analysis and ultrastructural localization studies had by then indicated that GP57/51 antigen and cruzipain were the same, or closely related molecular entities.[9] By extension, we deduced that GP25 antigen[36,37,38] corresponded to the CTE of cruzipain.[9] The finding that serum antibody reactivity of GP25 was sharply reduced upon periodate oxidation of the antigen and that serological reactions were inhibited by short O-linked oligosaccharides isolated from GP25 (Scharfstein and Mendonça-Previato, unpublished data) have suggested that the epitopes recognized by antibodies from chagasic patients were mostly formed by the carbohydrate moiety. Upon isoelectrofocusing,[38] we showed that GP25 in fact migrated as a heterogeneous set of 4 antigens of 25-kDa with pI's ranging from 4.38 to 5.30. Consistent with this, Hellman et al[31] found that the CTE was cleaved from the intact cruzipain molecule by an autocatalytic mechanism; they have further shown that the cleavage site, VVG/GPG, mapped to a proline rich segment that links the papain-like catalytic domain of the molecule to the CTE.

Progress in gene cloning[15] soon revealed that the 130 amino acid CTE consists of 3 distinct segments:

1. at the N-terminus, a protease sensitive hinge of 27 residues which includes 7 modified threonines and 7 proline residues

2. in the middle, a long 76 amino acid long "core", stabilized by 4 disulfide bridges, and containing one N-linked oligosacharide chain at Asn-255
3. a highly hydrophylic tail of 25 residues.

Out of the two putative N-glycosylation sites predicted for the catalytic domain, at least one is occupied by high-mannose type of oligosaccharides.[40] Given indications that the catalytic activity of purified GP57/51 was sensitive to temperature-dependent conformational changes,[24] it seemed plausible that the CTE could somehow modulate enzymatic function. Studies conducted with cruzain, the truncated recombinant enzyme, have recently ruled out this possibility.[27b]

In a different setting, Martinez et al[41] reported that chagasic serum antibodies reacted poorly with the natural (i.e., glycosylated) catalytic domain fragment (30-35-kDa). However, they noted that antibody binding to the intact natural cruzipain did not significantly disturb the enzyme's ability to hydrolyse small substrates. In an independent study, we have shown that formation of antigen-antibody complexes did not prevent natural protein inhibitors (e.g., cystatin C and H-kininogen[32]) from binding to the active-site cleft of cruzipain.[42] These data indicate that enzymatic function and antibody-binding activity of cruzipain are independently exerted by these two domains.[42,43,44]

A distinct distribution was found when Arnholdt and co-workers examined the domain distribution of T cell epitopes in cruzipain.[45] After conducting the screening of peripheral blood mononuclear cells from chronic chagasic patients with overlapping peptides spanning the full-length sequence of cruzipain, the immunodominant peptide specificities of $CD4^+$ T lymphocytes were mapped to a polymorphic segment of the catalytic domain (papain numbering of 177-209). The peptide-specific $CD4^+$ T cell lines isolated in this study were cross-stimulated with recombinant cruzipain and secreted IFN-γ, but not IL-4. Although more extensive studies are required to exclude the participation of peptides from the C-terminal domain in $CD4^+$ T lymphocyte activation during natural infection, these data point to an asymmetric distribution of T and B cell epitopes in the 345 amino acid-long mature cruzipain (Fig. 1). Additional studies are required to determine if the high degree of glycosylation and/or compactation of the CTE may somewhat reduce its sensitivity to proteolytic degradation by antigen-presenting cells (APCs). As discussed later, the CTE is probably less tightly folded in some subsets of cruzipain isoforms,[17,56] hence may yield antigenic peptides for $CD4^+$ T cells upon proteolytic degradation by lysosomal cathepsins. Similar constraints may influence susceptibility to proteosomal degradation, thus shaping the repertoire of cruzipain-specific $CD8^+$ T cells. Significantly, the latter were recently recognized as a relatively dominant specificity in intralesional $CD8^+$ T cells isolated in heart biopsies from HLA-$A2^+$ chagasic patients (Fonseca, S., Cunha-Neto, E., and Kalil, J., personal communication).

Targeting of the Amastigote Cruzipain with Synthetic Inhibitors

Early attempts to inactivate cruzipain during parasite intracellular development have exploited the differential specificity of irreversible peptidyl-diazomethyl ketone inhibitors,[28] some of which had been originally designed for mammalian cathepsin L. After verifying that Z-(SBz)Cys-Phe-CHN_2 efficiently inactivated purified cruzipain, we reported[28] that this membrane-permeable inhibitor could (i) partially block invasion of primary heart muscle cells by trypomastigotes, and (ii) arrest intracellular parasite development, when tested at the low μM range.[28] Significantly, we further showed that cruzipain was selectively targeted when the infected monolayers of heart cells were exposed to a closely related membrane-permeable inhibitor, Z-Phe-Tyr-CHN_2. These results for the first time suggested that the enzymatic activity of cruzipain was crucial for the intracellular development of the parasite.[28] These findings, along with those subsequently reported by Harth et al[46] and Franke de Cazzulo et al[47] have encouraged McKerrow and co-workers to explore cruzipain as a drug target.[48] After solving the three-dimensional structure of the catalytic domain of cruzain at high resolution,[21-22] new generations of pseudopeptide-based vinylsulfone inhibitors were designed, some of which have been shown to act as potent anti-parasite drugs when tested at therapeutic doses in animal models of Chagas disease.[49]

Structural Diversity of Cruzipain Isoforms

Although the available data suggest that cysteine proteinase activity is essential for parasite survival, technical obstacles involved in targeted deletion of the multicopy cruzipain genes[50] have thus far hampered the unequivocal demonstration of this concept. Another potential problem in assessing the functional significance of cruzipain is the existence of gene polymorphism[13,17,17b] that could possibly give rise to differential expression of isoenzymes at the various stages of the *T. cruzi* life cycle. Although the large majority of the cruzipain genes thus far sequenced encode for isoforms that are virtually identical to the cruzain archetype,[13,15,16,50] Lima et al have identified more divergent genes[17] after analysing transcripts obtained by RT-PCR. Sequence analysis of these DNA fragments revealed that in some molecules Glu-205 is replaced by uncharged residues, suggesting that the S2 pocket specificity of the corresponding enzymes may not possess the cathepsin B-like specificity displayed by natural[24,25] or recombinant cruzain.[16] In another divergent sequence, showing 86% amino acid identity to the cruzain gene (termed "cruzipain 2"), Glu-205 was preserved, although molecular modelling predicted the presence of non conserved amino acid substitutions close to the S2 subsite (Ser-61, Ser-68, Ser-69 and Ser-131, papain numbering) and to the S1'and S2' subsites (Fig. 1). Additional features distinguish the primary sequence of cruzipain 2 from those of the "cruzipain 1" complex.[17] For example, the first potential N-glycosylation site in the catalytic domain is absent from cruzipain 2, but there is an additional N-linked oligosaccharide chain predicted at the extreme end of the CTE (Fig. 1). Moreover, the last cysteine residue of the CTE of cruzain is not present in cruzipain 2. This suggests that out of the 4 pairs of disulfide bridges which form the CTE core of the major isoforms,[54] one is likely missing in cruzipain 2 (Fig. 1). Hence, it is possible that the CTE of cruzipain 2 is not as tightly folded as in the cruzain-type of proteins.

In order to verify if gene polymorphism has contributed to substrate specificity, we have cloned a cruzipain 2 gene and expressed it as an active enzyme[51] in *S. cerevisiae*. Analysis of the kinetic properties of the recombinant cruzipain 2, which was also truncated in the CTE, has revealed that it differs from either cruzain or natural cruzipain, with respect to substrate preference as well as to susceptibility to inactivation by synthetic inhibitors.[51] Assays performed with substrates showing modifications on P1-P3 and P1'-P3' have recently indicated that cruzipain 2 has a more narrow substrate specificity as compared to cruzain or the natural enzyme purified from Dm28c parasites (Reis et al, in preparation). Using a combination of biochemical and kinetic methods, Lima and co-workers[51] have recently demonstrated that members of the cruzipain 2 subset are expressed at higher levels by amastigotes and trypomastigotes. Of further interest, we recently observed[11] that trypomastigotes overexpressing cruzipain 2 invade a broader range of mammalian cells as compared to parasites that overexpress full-length cruzipain 1. Interestingly, the anti-parasite drug Mu-F-hf-VSPh, which is a potent inhibitor of cruzain,[49] is also able to inactivate recombinant cruzipain 2, albeit at reduced efficiency (Lima, APL and Scharfstein, J., unpublished data). In light of these findings, and of evidence that drug resistant cell lines emerge when epimastigotes are subjected to selective pressure with different synthetic inhibitors of cruzipain,[52,53] the long-term term benefits of Mu-F-hf-VSPh treatment in vivo are worth evaluating.

Cazzulo, Parodi and co-workers[54-56] have described a second biochemical mechanism underlying the structural variability of cruzipain isoforms. After realizing that purified cruzipain from the Tul2 strain separates as 12 distinct bands upon isolectrofocussing,[32] they have verified that this heterogeneity stems, at least in part, from changes in a single N-linked carbohydrate moiety (Asn-255) in the C-terminal domain. These consist either of high mannose, hybrid mono-antenary or complex bi-antenary oligosacharide structures.[55] The genetic basis for this biochemical heterogeneity was then investigated by RT-PCR. Among the 48 cDNA clones which they isolated, 8 were found to correspond to structural variants in the C-terminal domain.[56] It was also noticed that some of these nonconservative substitutions may account for the charge heterogeneity that has been observed in the purified CTE [54] and likewise, in GP25 antigen.[38] As pointed out by Cazzulo and co-workers,[54,56] the presence of charged amino acids may induce subtle changes in the three-dimensional structure of the CTE. This may then affect the degree of oligosaccharide chain exposure to the processing enzymes that generate these variable carbohydrate structures in distinct cruzipain isoforms. Similar mechanisms may be responsible for the elusive structural changes that affect the threonine cluster[54] that links the CTE to the catalytic domain (Fig. 1). Although the functional consequences of N-linked oligosaccharide variability in cruzipain molecules has not been evaluated, this may account for differential enzyme stability,[24,24b] variable antigenic properties[36-38,41,41b,43] and/or subcellular localization.[9,57-59]

Regulation of Cruzipain Activity during Parasite Development

Details about the structural organization of the cruzipain gene family from the Sylvio X10.6 strain emerged from analysis of clones isolated from cosmid libraries.[50] These studies revealed presence of two allelic gene clusters composed of approximately 14 and 23 tandemly repeated genes as well as genes that could be dispersed copies of closely related cysteine-proteinase genes. At the 3'-end of each tandem array, a divergent sequence with the potential to encode an active enzyme with a short (49 amino acid) hydrophobic CTE was identified, but the relationship, if any, of these putative enzymes with the partially characterized membrane-bound forms of cysteine proteases that have been reported by different groups[58,59] has not been clarified.

Although epimastigotes are obviously capable of upregulating lysosomal proteolysis to satisfy their stringent metabolic needs at times of cellular division, overexpression of cruzipain in axenic parasites has enhanced metacyclogenesis.[6] It is intriguing that cruzipain is sensitive to substrate inhibition[24,24b,27b] because the modulation is temperature-dependent, being manifested at ambient temperatures prevailing in the insect host, but not at 37°C.[24,24b] It will be interesting to know if substrate inhibition occurs during *T. cruzi* development in the triatomine gut, perhaps reflecting an adaptive response to changes in the content of blood proteins ingested by the insect.

Analysis of the stage-regulated expression of cruzipain has indicated that comparable levels of RNA are produced by the various developmental forms of the parasite,[50] although measurements of protein levels revealed that epimastigotes consistently display higher amounts as compared to trypomastigotes and amastigotes.[16,47,50,61] Due to the inherent diversity of cruzipain isoforms, and to the fact that cathepsin B-like cysteine proteinases are also expressed by these parasites,[53,64] estimates of protein and activity levels have in the past generally encompassed a broad group of homologous proteinases operationally referred to as the major lysosomal cysteine proteinase.[54] In some studies, the enzyme activity of cruzipain in epimastigotes was found to be 10-100 fold higher than in trypomastigotes and amastigotes.[54] However, recent studies revealed that the mammalian stages of the life cycle accumulate high levels of "chagasin", a tight binding and structurally unique endogenous inhibitor of papain-like cysteine proteinases.[62] Given that chagasin is present in low amounts in epimastigotes,[62] the cysteine protease activity measurements in lysates from trypomastigotes and amastigotes may have been underestimated. Considering that the lysosomal cruzipain represents 3-4% of the soluble protein of epimastigotes,[63] the upregulation of chagasin expression by trypomastigotes and amastigotes may contribute to the poorly understood post-translational mechanism(s) of cysteine protease regulation.[50]

The Cell Surface Expression of Cruzipain Molecules Is Developmentally Regulated

In spite of recent advances in the characterization of cysteine protease trafficking pathways in trypanosomatids,[19,65,66] the mechanisms responsible for the developmental regulation of cruzipain surface expression remain poorly characterized. Early attempts to radioiodinate surface molecules demonstrated the presence of a cruzipain product of lower molecular mass (i.e., GP51) at the surface of epimastigotes, but not on bloodstream trypomastigotes.[38] Analysis of metacyclogenesis has on the other hand revealed that the cell surface display of these cruzipain-like antigens on epimastigotes is markedly reduced when the dividing forms transform into metacyclic trypomastigotes.[67] Surface display of active cysteine proteinases at the cell surface of amastigotes was demonstrated by probing infected host cells in culture with active site-directed biotinylated inhibitors.[49] Treatment of freshly released amastigotes with cystatin-like inhibitors (e.g., biotin-LVG-CHN_2) also yielded intense surface labelling by confocal microscopy whereas only faint reactions were observed on trypomastigotes (Fig. 2A). Consistent with this differential staining, gold-labelled polyclonal[66b] or monoclonal antibodies (Fig. 2B) revealed presence of cruzipain-like antigens at the cell surface of amastigotes, but the monoclonal antibodies reacted poorly with the surface of trypomastigotes (Fig. 2C). As described in early ultrastructural studies,[9] there is a preferential accumulation of cruzipain antigens in endolysosomal-like vesicles and in the flagellar pocket of trypomastigotes (Fig. 2C). By preventing the exposure of a large number of these highly antigenic proteases at the cell surface of trypomastigotes, these extracellular parasites may be spared from potentially harmful attack of anti-GP25 antibodies, present at high titers in chronic chagasic serum.[36,37,41] Protected by the intracellular environment, not only are amastigotes relieved from the threat of antibodies to cruzipain, but the cell surface localization of the proteinase may facilitate the breakdown of host cytoplasmic proteins, some of which may act as precursors for mitogenic peptides and/or signals required for terminal differentiation.[28] The evidence that membrane-permeable synthetic cysteine protease inhibitors arrest intracellular growth of amastigotes offers circumstantial support to this hypothesis.[28,46,47] However, it is still unclear if these inhibitors may impair lysosomal function and/or provoke lesions of Golgi complex cisternae of the amastigotes, as described for axenic epimastigotes.[19]

In spite of the lack of a precise understanding of the cruzipain function in amastigotes, it is possible that the activity of this protease is modulated by metabolic changes that occur in the infected host. For example, it will be interesting to know if the protection which activation of the NFk-B pathway affords to particular cell types[4] stems from regulatory constraints on the cruzipain activity expressed by amastigotes. This possibility is worth exploring in light of a recent report by Venturini ands co-workers[68] showing that purified cruzipain is inhibited by nitric oxide (NO), presumably due to S-nitrosylation of the Cys-25 active site. Along similar lines, the efficacy of cruzipain inactivation may vary from one type of infected host cell to another due to fluctuations of type I cystatins (e.g., stefins A/B), a potent class of protein inhibitors of papain-like cysteine proteinases. Of potential relevance to pathogenesis are the recent findings[51] that cruzipain 2 is significantly less sensitive to inactivation by tight-binding natural proteinase inhibitors from the cystatin superfamily, such as cystatin C and H-kininogen. Thus, it is conceivable that *T. cruzi* may have benefited from structural diversification of cruzipain active sites to reduce the efficiency of inhibitory interactions mediated by different cystatins, at various stages of parasite development. As discussed later in this text, this concept has provided

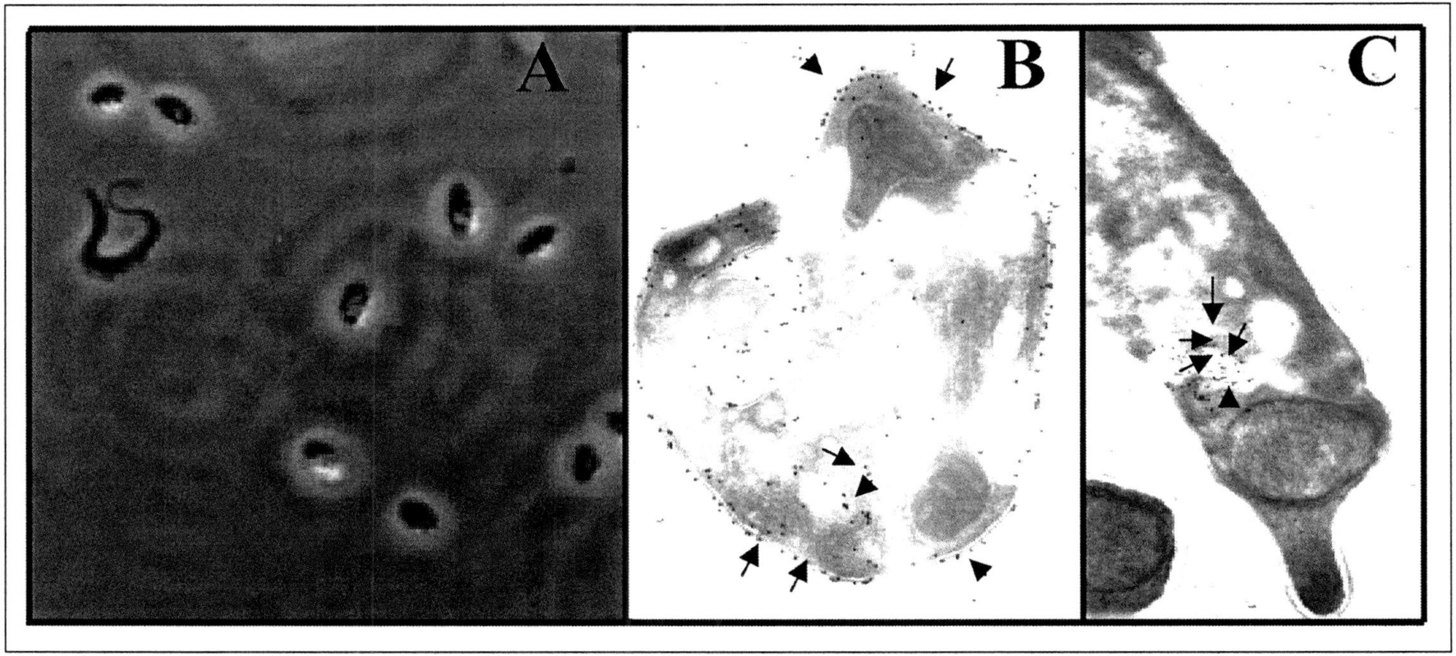

Figure 2. The surface expression of cruzipain is regulated during parasite development. Left panel, tissue-culture derived trypomastigotes (TCT) and amastigotes (Dm28c) analysed by confocal microscopy , following treatment of parasite suspensions with an irreversible cruzipain inhibitor, biotin-LVG-CHN_2.[53,112] The freshly released parasites were recovered from the supernatants of monolayers of overinfected Vero cells. After fixation, the cells were treated with streptavidin-FITC. Parasite labelling with biotin-LVG-CHN_2 was completely abolished by adding excess of E-64 to the culture medium. Note the intense surface labelling of extracellular amastigotes while trypomastigotes only exhibit a discrete intracellular labelling (only observed at high magnifications). Middle panel. Ultrastructural immunocytochemistry performed with monoclonal antibodies (mAb 212BH6) to cruzipain showed conspicuous labelling of the cell surface of amastigotes, as well as presence of intracellular gold particles (N. Andrews and J. Scharfstein, unpublished). Right Panel. Treatment of TCT with the mAb anti-cruzipain (212BH6) revealed many particles in the flagellar pocket and endolysosomal-like vesicles while surface exposure of this cruzipain epitope is occasionally observed.[9]

a framework for studying the interplay between the extracellular activities of cruzipain and host kininogens, a type III cystatin, prior to host cell invasion.[10b,11,113]

Cruzipain Diversity in Amastigotes: Possible Implications to Immunopathology

Using monoclonal antibodies to probe heart autopsies from chronic chagasic patients, we have found prominent antigen deposits of cruzipain amidst the myocardial inflammatory infiltrates.[72] Although the origin of these antigenic aggregates was not determined, they most likely result from amastigote remnants, rather than intracellular pseudocysts, which are rarely detected in tissues from chronically infected patients. As we have recently proposed,[71] high numbers of amastigotes may accumulate in extracellular spaces when the intracellular life cycle of *T. cruzi* is prematurely terminated, either because of early immune attack by antigen-specific class I-MHC restricted $CD8^+$ CTLs or due to premature activation of a pro-apoptotic pathway. Unlike trypomastigotes, which have a long flagellum and are able to migrate from the primary site of infection, the amastigotes tend to persist in the inflammatory foci, thus serving as a potent source of pro-inflammatory factors[113] and antigens.[71,72] In a provocative biochemical study, Stoka et al[69] have recently shown that purified cruzipain activates, albeit at low efficiency, mammalian pro-caspases 3 and 7 in vitro. These authors speculated that genes encoding pro-apoptotic cysteine proteinases could have been positively selected by an intracellular pathogen, such as *T. cruzi*, since lysosomal cysteine-proteases from mammalian cells, including the closely related cathepsin L, do not display pro-caspase activity. Of potential interest in this context, Freire-de-Lima and co-workers have recently demonstrated that TGF-β stimulates the intracellular growth of *T. cruzi*[70] by inducing macrophage suppression through apoptotic bodies, which in their study, have originated from dead lymphocytes. Although appealing, the possibility that intracellular parasites may rely on cruzipain to trigger apoptosis in infected host cells is remote because this particular protease (i) activates pro-caspase 3/7 very inefficiently, and (ii) is expressed at high levels by the dividing cells, i.e., amastigotes. Furthermore, activation of a pro-apoptotic pathway by cruzipain is unlikely because it would probably kill the host cells before the amastigotes could undergo transformation into trypomastigotes, thus giving rise to abortive cycles of infection.[71] Although speculative, it is possible that at late stages of their life cycle, the intracellular trypomastigotes may trigger host cell apoptosis by expressing subsets of cruzipain isoforms that may act as efficient pro-caspases in some mammalian cells. This possibility deserves to be explored in the near future.

Antigen-Presentation of Cruzipain: A Role for α_2-Macroglobulin Receptor (CD91)

Unlike their mammalian counterparts, cathepsin B and L, cruzipain isoforms are enzymatically stable and active over a wide range of pH (5-7.5). Whether originating from amastigotes or from trypomastigotes, extracellular cruzipain is probably rapidly inactivated due to oxidation of the reactive site cysteine residue or by plasma-borne proteinase inhibitors (eg, α_2-macroglobulin or kininogens). Our early studies focused on α_2-macroglobulin, because it has been shown that this broad-spectrum plasma proteinase inhibitor can modulate *T. cruzi* activity in a variety of settings.[73,74] The existence of a close spatial association between parasite antigen and the distribution of α_2-macroglobulin[75] in the myocardium of infected mice, together with evidence for recruitment of macrophage-like cells expressing the α_2-macroglobulin scavenger receptor (α_2MR/CD91) into sites of myocardial inflammation in the human heart[72] suggested that interactions with the proteinase inhibitor may ultimately promote the clearance of the parasite proteinase from interstitial spaces. After showing that α_2-macroglobulin could indeed entrap a fraction of purified cruzipain molecules in vitro, Morrot et al[72] have demonstrated that α_2-macroglobulin-cruzipain complexes were rapidly internalized by α_2MR/CD91 from human monocytes. As predicted, the endocytic uptake of these antigen-complexes by the CD91 scavenger receptor resulted in highly efficient processing and presentation of cruzipain-derived peptides to class II MHC-restricted $CD4^+$ T cells from chagasic patients, a process that drastically reduced the activation threshold of the lymphocytes by the cruzipain antigen.[72] Recent analysis of the priming of $CD8^+$ T cells by antigens internalized by dendritic cells and macrophages indicated a key role for CD91 receptor in MHC class I pathway of antigen presentation.[72b] At least for the peptides chaperoned by heat shock proteins and internalized by CD91 it was clearly shown that post-uptake processing involves proteasomes and the transporters, and hence involves the conventional "endogenous" antigen processing route. Predictably, the CD91-mediated uptake of cruzipain/α_2-macroglobulin may likewise favor the priming and/or secondary stimulation of cruzipain-specific $CD8^+$ T cells. Consistent with a role for CD91 in adaptive immunity,[72] Araujo-Jorge et al[76] have recently reported that infected mice with targeted deletion of α_2-macroglobulin genes developed an exacerbated form of myocarditis. They further showed that levels of serum antibodies against the parasites were substantially decreased in the knockout mice, as compared to wild-type animals.[76] Interestingly, Giordanego et al[41b] recently found cardiac autoreactive antibodies in mice immunized with cruzipain (devoid of enzymatic activity) and proposed that myosin and cruzipain share cross-reactive epitopes. The structure of this putative cross-reactive myosin/cruzipain epitope must be characterized, in light of indications that vaccination with recombinant cruzipain protect mice from systemic and mucosal challenge with *T. Cruzi* systemically and mucosally.[76b,76c] By contrast to the IL12-driven immunity afforded by experimental vaccines, immunization with the purified enzyme (i.e., without adjuvant) preferentially induces T2 type immunity,[76d] hence may ultimately favor intracellar parasite growth in macrophages. Collectively, these studies suggest that CD91-dependent uptake of α_2-macroglobulin/cruzipain complexes may contribute to adaptive immunity and/or immunopathology (Fig. 3).

Activation of Kinin-Receptors Potentiates Host Cell Invasion by Trypomastigotes

For many years, the evidence linking trypomastigote infectivity to the activity of cysteine proteases was at best circumstantial. Invasion assays performed in different culture systems suggested that polyclonal antibodies could protect host cells from infection. Intriguingly, antibodies exclusively directed against GP25 partially protected primary cultures of smooth muscle human cells from trypomastigote invasion, while failing to reduce parasite infectivity of primary cultures of human fibroblasts isolated from the same individual.[77] In an independent study, Souto-Padron et al[57] noted that polyclonal antibodies raised against native cruzipain protected macrophages from invasion. Uncertainties concerning the fine specificity of the polyclonal antibodies used in these early studies[57,77] precluded a definitive assessment of the molecular mechanism underlying their protective effects.

Evidence indicating that proteases play a crucial role in the invasion process can be traced back to the pioneering work by Piras and co-workers,[78] who reported on the effects which serum "priming" imparted on the phenotypic maturation of tissue culture trypomastigotes. Work along similar lines indicated that the proteinase inhibitor α_2-macroglobulin reduced macrophage uptake of bloodstream trypomastigotes.[73,74] Our findings that synthetic inhibitors directed to the cathepsin L-like cruzipain partially protects primary cultures of mouse heart cells from invasion by tissue culture trypomastigotes[28] suggested, for the first time, that the catalytic activity of cruzipain could possibly be involved in this process.

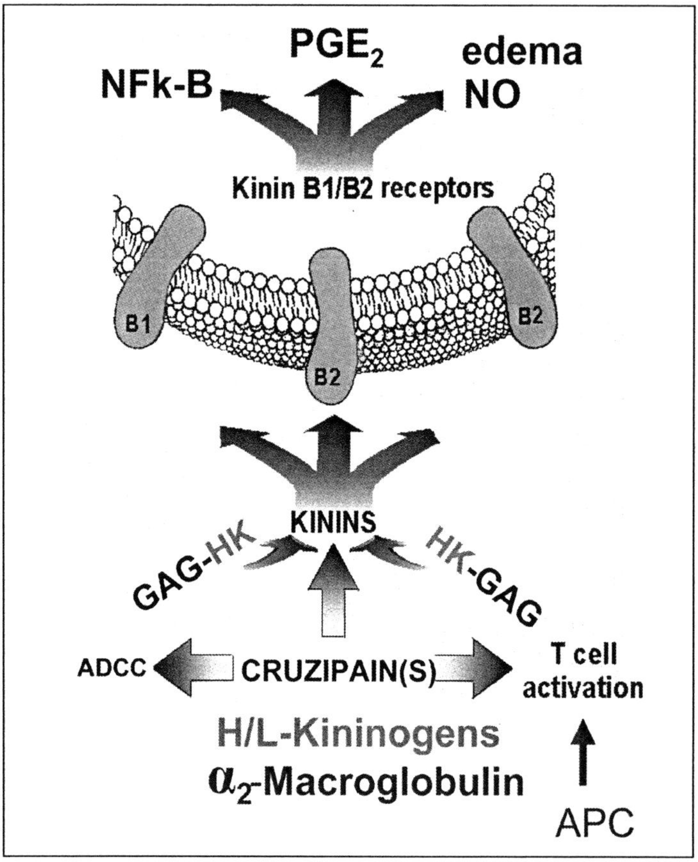

Figure 3. Cruzipain Interaction with Host Proteinase Inhibitors: Role in Parasite-Induced Immunopathology. The scheme illustrates the molecular pathways by which cruzipain may induce pathology. Antigenic deposits of cruzipain in the myocardium of chagasic patients[72] are thought to derive mainly from extracellular amastigotes.[71] Interactions between host protease inhibitors (eg. α2-macroglobulin and the kininogens) and catalytically active forms of cruzipain[72] may likely occur as soon as plasma proteins leak into the site of infection. Cruzipain then releases vasoactive kinins from kininogens in a reaction positively modulated by heparan sulphate proteoglycans. [10b] This event results in the activation of the endothelium and/or smooth muscle through constitutive (B2)[11] or inducible (B1) kinin receptors.[113] Responses transduced through the kinin receptors increase host cell susceptibility to parasite infection, [11] induce plasma leakage,[113] stimulate NO release and/or enhance prostaglandin biosynthesis.[80] Our recent demonstration that kinin peptides activate immature dendritic (DCs) through B2-kinin receptors[115] may ultimately link cruzipain enzymatic activity to innate immune responses[76d] evoked by the parasites. Interactions between cruzipain and α2-macroglobulin may indirectly modulate cellular immune responses.[72,75-76d] because antigen-presenting cells, eg. macrophages and dendritic cells, rely on a highly endocytic scavenger receptor (CD91/α2-macroglobulin receptor)[72b] to internalize cruzipain/α2-macroglobulin complexes.[72] The threshold for the activation of antigen-specific $CD4^+$ T cells is markedly reduced due to enhanced presentation of cruzipain peptides by the MHC class II pathway.[72] In other settings, it has been shown that CD91-dependent uptake of cruzipain/α2-macroglobulin complexes by dendritic cells[72b] may also contribute to the priming and/or secondary stimulation of MHC class I restricted $CD8^+$ T cells.[72b] In addition to the inhibitory effects on proteinases, α2-macroglobulin attenuates tissue damage induced by inflammatory cytokines (eg. TNF-β).[73,75,76] Upon endothelial cell injury, kininogens docked to sulphated proteoglycans may bind soluble cruzipain molecules through accessible cystatin domains.[42] This may allow for aggravation of microangiopathy due to antibody-dependent cellular cytotoxicity (ADCC)[97] due to recognition of immunodominant CTE epitopes. [36,37,41,42]

The notion that cruzipain could act as a kinin-releasing enzyme *in vitro*[10] and awareness that kinin peptides stimulate IP_3-mediated $[Ca^{2+}]_i$ responses in mammalian cells expressing kinin-receptors[79] has motivated attempts to investigate the involvement of this signalling pathway in *T. cruzi* invasion. The term "kinin" refers to a small group of vasoactive metabolites structurally related to the nonapeptide bradykinin. Although the kinin peptides have been traditionally viewed as classical mediators of acute inflammation (e.g., inducers of oedema formation, vasodilation and pain sensations),[80] progress in vascular research has more recently characterized the kinin system as a more general modulator of circulatory homeostasis.[81] In order to explore the mechanisms by which the tightly controlled pro-inflammatory kinin cascade is activated by *T. cruzi*, the molecular components of this system must be briefly described. Generation of these paracrine vasoactive mediators depends on the proteolytic excision of the internal kinin segment from their plasma precursor proteins, high or low molecular weight kininogens (H- and L-kininogen), either by plasma or tissue kallikrein. Upon vascular injury, the activated forms of plasma kallikrein promote the release of bradykinin from H-kininogen.[82] However, during inflammation, H-and L-kininogens extravasate from blood capillaries to extravascular tissues, where they suffer proteolytic attack by tissue kallikrein (Fig. 4,A), liberating lysyl-bradykin, i.e., "kallidin".[80] In other pathological settings, leukocyte proteases with a less restricted specificity may liberate kinins by processing their flanking sites in kininogen cooperatively. For example, the concerted action of neutrophil elastase and mast cell tryptase on oxidized forms of kininogens may also liberate a slightly larger kinin, Met-Lys-bradykinin.[83] Once released, the short-lived kinins (half life of <15 sec in the plasma) exert their biological effects by the paracrine mode, through the activation of distinct sub-types of heterotrimeric GPCRs, B2 and B1 (Fig. 4B). Long-range effects on bradykinin receptors localized at the vascular lining are prevented by the metabolic action of kinin-degrading peptidases[80] (Fig. 4A), e.g., the angiotensin converting enzyme (ACE), also termed as kininase II. While intact kinins (BK or lysyl-BK, i.e., kallidin) are the agonists for the constitutively expressed B2-bradykinin receptor, the B1 receptors, which are upregulated during inflammation[84] are preferentially activated by kinins that are deprived of the C-terminal Arg by carboxypeptidase N/M, i.e., kininase I (Fig. 4A).

Generated by differential RNA splicing of a single gene, the H- and L-kininogens are prototypes of modular proteins.[85] Before discussing how cruzipain releases kinin peptides from kininogens, the domain organization of these plasma glycoproteins must be reviewed (Fig. 4A). Starting from the N-terminus, the first 4 domains (D1-4) are shared by H- and L-kininogen. D1-3 domains are homologous to cystatins, but only the last two (D2-3) are functionally active inhibitors of cysteine proteinases. Adjacent to D3 is the bradykinin sequence (D4). At the C-terminal side of the bradykinin moiety of H-kininogen, there are two unique domains (D5H and D6H) that mediate contact phase activation of the coagulation cascade. Of possible relevance to the kinin-releasing reaction mediated by cruzipain, the binding sites which mediate the binding of kininogens to endothelial cells also map to D3 and hence overlap with the last cystatin domain.[86,87] Docking of H-kininogen to endothelial cell surfaces is further stabilized by interactions mediated by the histidine-rich sequence in D5H with heparan sulphate proteoglycans.[87] In other cell types (e.g., smooth muscle), H-kininogen is tethered to the cell surfaces by binding to chondroitin sulphate proteoglycans.[87b]

At first sight, the realization that kininogens (type 3 cystatins) could serve as a natural substrate for cruzipain seemed paradoxical, because soluble forms of kininogens act as potent inhibitors of most papain-like cysteine-proteinases, including cruzipain.[32,42,51] We reasoned, however, that H-kininogen may not display functionally active cystatin-like domains once it docks to the endothelial cells, because some of the sites involved in cell surface binding have been mapped to its D3 domain,[86,87] hence overlap with the last cystatin-like structure (Fig. 5A). Consistent with this notion, fluorescein-labelled cruzipain did not bind appreciably to the

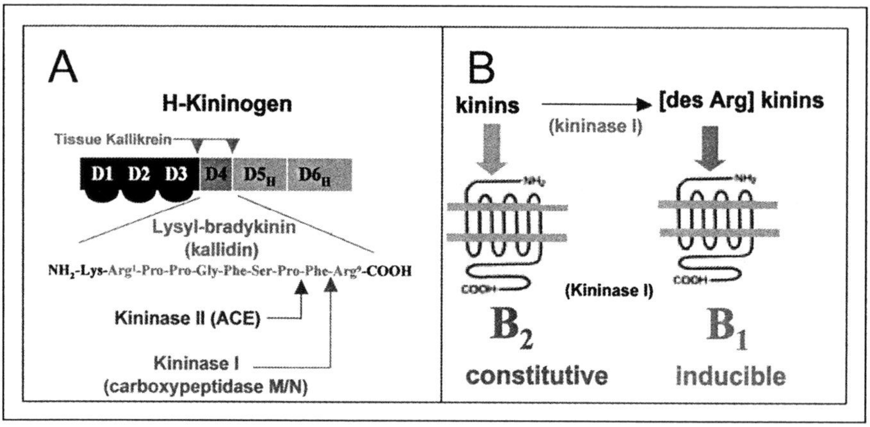

Figure 4. Kinin activation pathways. The scheme shown in panel A illustrates the multi-domain organization of one of the kinin-precursor glycoproteins, H-kininogen.[85] The internal bradykinin moiety is located in the D4 domain. H-kininogen and L-kininogen (not shown) share the cystatin-like domains (D1-3) and the adjacent kinin moiety. Tissue kallikrein or plasma kallikrein cleave the C-terminal flanking side of the kinin peptide at the same bond. However, the N-terminal flanking bonds of the kinin segment[80] is not cleaved at the same position by these serine proteases: human plasma kallikrein (not shown) releases a slightly shorter fragment from H-kininogen, the nonapeptide bradykinin, while tissue kallikrein cleaves the Met-Lys bond, thus liberating the decapeptide Lys-bradykinin (kallidin) from either H- or L-kininogen. Note that optimal triggering of the B2 kinin receptor subtype depends on the presence of the C-terminal Arg in bradykinin or kallidin. The B2 kinin receptor subtype is constitutively expressed at high levels by endothelial and vascular smooth muscle cells[80] in vivo (panel B). The cleavage specificity of two main kinin-degrading peptidases is also indicated in the scheme. Kininase II, a di-peptidyl carboxypeptidase (also known as ACE, the angiotensin converting enzyme) swiftly inactivates the vasoactive and pro-inflammatory activity of the kinin peptides (panel A). During inflammatory states, the expression of the B1 kinin receptor subtype is upregulated by several cell types,[84] such as vascular endothelial cells, macrophages and fibroblasts (panel B). Signalling of the inducible B1 kinin receptor is conveyed by metabolites (des-Arg^{9}-bradykinin or des-Arg^{10}-kallidin) generated by the processing action of kininase I (eg. carboxypeptidases M/N) respectively on bradykinin or kallidin[84] (panel A).

cell surface of primary cultures of human umbilical vein endothelial cells (HUVECs), suggesting that the cystatin domains from cell-bound forms of kininogens are not available for tight-binding interactions with soluble forms of the parasite protease (A. Morrot and J. Scharfstein, unpublished data). Consistent with this view, vigorous $[Ca^{2+}]_i$ influx was observed when purified cruzipain (pre-activated) was added to monolayers of Chinese Hamster Ovary cells transfected with the rat-B2 bradykinin receptor gene (CHO-B2).[11] The data reproduced in Figure 5B show that the $[Ca^{2+}]_i$ response induced by purified cruzipain was blocked by HOE 140, a specific antagonist of the B2 kinin-receptor subtype, or by E-64, an irreversible inhibitor of papain-like cysteine proteinases, hence confirming that cruzipain has competence to liberate the kinin agonist from cell-bound forms of the kininogens.[11] These findings suggested that the bradykinin segment in cell-bound kininogens could be "presented" to the trypomastigotes in a conformation that is favorable for proteolytic excision by cruzipain. We then tested this prediction by monitoring $[Ca^{2+}]_i$ transients in CHO-B2 following addition of living parasites.[11] Our data (Fig. 6A) showed that Dm28 trypomastigotes vigorously trigger repetitive $[Ca^{2+}]_i$ transients in CHO-B2 but not in mock-transfected CHO cells.[11] Importantly, the intensity of the $[Ca^{2+}]_i$ responses conveyed by the B2-bradykinin receptor correlated with increased susceptibility to parasite invasion[11] (Fig. 6B). Importantly, addition of membrane-permeable cruzipain inhibitors to the cultures significantly reduced these effects.[11]

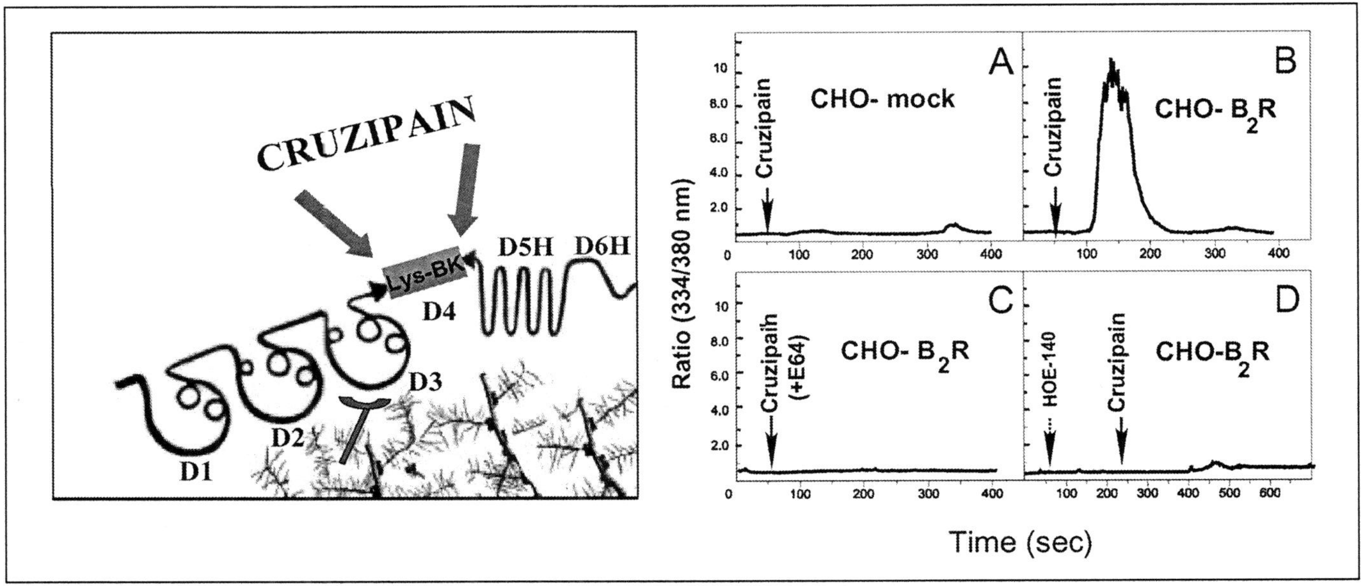

Figure 5. Cruzipain releases kinins by acting on surface-bound kininogens and triggers vigorous $[Ca^{2+}]_i$ in host cells that express the B2 kinin receptor. Left panel. Schematic representation of interactions between cruzipain, H-kininogen and heparan sulphate proteoglycans at the surface of endothelial cells. Searches for the endothelial docking sites for H-kininogen [86] have recently converged on heparan sulphate proteoglycans[87] while chondroitin sulphate undertakes this tethering function in other cell types.[87b] Analysis of the structural basis of H-kininogen interaction with heparan sulphate indicated that the proteoglycans' negatively charged chains bind to the histidine-rich sequence ($D5_H$) of H-kininogen.[87] A second binding site(s) for heparan sulphate overlaps with the cystatin-like (D3) segments, hence may account for the drastic loss in the potency of cruzipain inhibitory activity of H-kininogen.[10b,32,42] N-terminal sequence analysis of the H-kininogen's breakdown products revealed that the heparan sulphate chain re-directs the cleavage specificity of cruzipain.[10b] These cooperative effects are thought to facilitate the release of the kinin moiety by cruzipain.[10b] Right Panel. Data reproduced from Scharfstein et al[11] showing that cruzipain triggers vigorous $[Ca^{2+}]_i$ responses in CHO cells overexpressing the B2-bradykinin receptor (CHO-B2) but not in CHO-mock cells. Serum-borne kininogen molecules tend to accumulate at the cell surface of these host cells. $[Ca^{2+}]_i$ responses were determined by adding 5 nM of the activated peptide to Fura 2-AM treated cells that were briefly cultivated in serum-free DMEM medium supplemented with 1 mg/ml BSA and 25 μM of captopril (ACE/kininase II inhibitor).

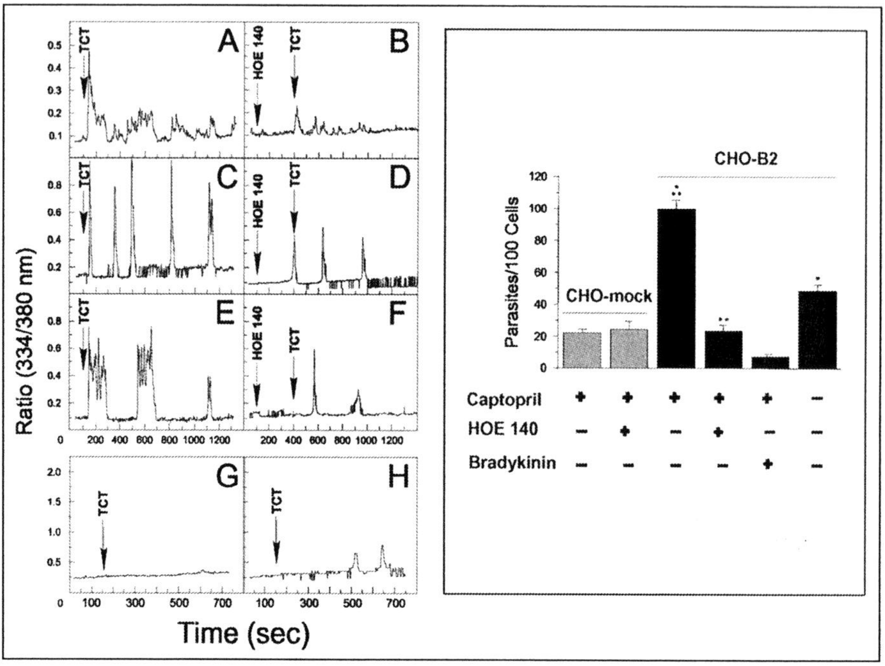

Figure 6. Activation of B2-bradykinin receptors triggers $[Ca^{2+}]_i$ and potentiates host cell invasion by Dm28 trypomastigotes. Left panel, tracings reproduced from Scharfstein et al[11] represent $[Ca^{2+}]_i$ transients yielded by Fura-2 treated CHO cells (single cells or average measurements) upon exposure to TCT (Dm28c). The results show that the parasites trigger repetitive $[Ca^{2+}]_i$ responses in CHO-B2 but not in CHO-mock. Addition of a specific antagonist of B2-kinin receptor (HOE 140, 100 nM) to cultures of CHO-B2 prevented the $[Ca^{2+}]_i$ responses. Cellular interactions were carried out in serum-free medium supplemented with BSA and captopril. Similar observations were made with primary cultures of human umbilical endothelial cells. Right Panel, the data represent the number of intracellular parasites observed after 3 h incubation, using the same conditions described above.[11] Note that the ACE inhibitor increases host cell susceptibility to parasite invasion, but only in cultures containing CHO-B2. Similar to the $[Ca^{2+}]_i$ responses, these effects were blocked by HOE 140 or by excess of bradykinin (due to down-regulation of the B2-kinin receptor). Invasion assays with HUVECs have yielded similar results.[11]

Similar results were obtained with host cells that naturally express B2 subtype of kinin receptors such as HUVECs.[11]

As expected from this paracrine signalling system, the efficiency of trypomastigote invasion by the kinin-receptor transducing pathway was subtly regulated by the activity of kinin-degrading peptidases.[11,113] The data reproduced in Figure 6 (panel B) indicate that captopril, a potent inhibitor of the kinin-degrading peptidase (ACE/kininase II), drastically increased (over 4-fold) parasite invasion of CHO-B2 or HUVECs, but not of CHO-mock. Moreover, addition of monoclonal antibodies to the bradykinin epitope of kininogens reduced invasion to background levels, thus confirming that trypomastigotes were capable of releasing the kinin peptide moiety from cell-bound kininogens. Consistent with these results, engagement of the kinin-receptor signalling pathway was reduced in serum-free medium. However, parasite infectivity was restored when we added purified H-kininogen or alternatively, an exogenous source of the peptide agonist, i.e., bradykinin, to the serum-free medium.[11] Interestingly, the infection peaked at ~10 nM of this agonist while high concentrations of bradykinin (~100 nM) protected the cells from parasite invasion.[11] The inhibition observed at high doses of the bradykinin agonist is caused by down-regulation of the B2-kinin receptor,[81] a mechanism that may prevent host cells from being overinfected when simultaneously exposed to a high number of trypomastigotes. In summary, our studies indicated that the outcome of the host-parasite interaction is delicately modulated by levels of the kinin–precursor molecules (i.e., kininogens) and by the activity of kinin-degrading peptidases (e.g., ACE).

The Kinin-Releasing Activity of Trypomastigotes Is Linked to Cruzipain 1

The possibility that the kinin-releasing activity of trypomastigotes is influenced by the repertoire of cruzipain isoforms expressed by any given parasite clone has only been superficially investigated thus far. Previous studies with fluorogenic substrates[10] have indicated that the bradykinin flanking site sequences are hydrolysed at different efficiencies by cruzipain 2 and cruzain (designated hereafter as cruzipain 1). Given that both isoforms are detected in supernatants obtained by short-term cultivation of trypomastigotes, we have evaluated their contribution to cellular invasion by overexpression of full-length cruzipain 1 and cruzipain 2 genes in transfected parasites.[11] Our data, although indirect, supported a causal link between enhanced expression of cruzipain-1 and the competence to invade cells via the kinin signalling route. Perhaps not surprisingly for a genetically diverse species such as *T. cruzi*, we have recently observed that the kinin-activating phenotype of tissue culture trypomastigotes is not ubiquitously distributed in strains/clones routinely used in laboratory studies (Moreira da Silva et al, unpublished). Preliminary analysis suggests that these parasite strains secrete cysteine proteases at similar levels to Dm28c trypomastigotes. However, the proteases isolated from some of these parasite strains were not able to liberate kinins from purified H-kininogen at high efficiency, suggesting that the substrate specificity is not the same as cruzipain-1. These results are in agreement with the proposition that the parasites' differential ability to activate the kinin cascade may be linked to the expression of particular subsets of cruzipain isoforms.[11]

Kinin-Release by Trypomastigotes Is Enhanced by Cooperative Interactions between Heparan Sulphate, H-Kininogen and Cruzipain

Given indications that glycosaminoglycans (GAGs) serve as platforms for the cell-surface accumulation of H-kininogen in a variety of mammalian cells[87,87b] we have recently investigated if interactions with proteoglycan chains affected the kinin-releasing activity of cruzipain.[10b] N-terminal sequence analysis of the multiple breakdown products of H-kininogen revealed that heparan sulphate re-directs the substrate specificity of cruzipain.[10b] Interestingly, peptide bonds localized in the cystatin-like segments of H-kininogen were more efficiently hydrolysed when we added this GAG to the reaction mixture. Consistent with these observations, the cysteine inhibitory activity of H-kininogen was drastically reduced (~10 fold increase of K_{iapp}) in samples supplemented with heparan sulphate. Similar effects were observed when we assayed cruzipain with a fluorogenic peptide substrate spanning the N-terminal flanking side of bradykinin. Kinetic data indicated that the catalytic efficiency of cruzipain is enhanced up to 6-fold in the presence of heparan sulphate. We then tested if the multiple effects ascribed to this GAG would translate into a more efficient kinin-releasing reaction. Optimal responses were indeed obtained, but only at relatively narrow stoichiometry ratios of cruzipain, H-kininogen and heparan sulphate.[10b] Significantly, addition of the GAG to suspensions of trypomastigotes resulted in ~35 fold enhancement of kinin release.[10b] Collectively, these studies have demonstrated that cooperative interactions between cruzipain, HK and heparan sulphate proteoglycans may significantly enhance the kinin-releasing activity of trypomastigotes.

Intercellular Spaces May Act As Privileged Sites for the Kinin-Release of Cruzipain

As discussed earlier in this review, natural or synthetic inhibitors of cysteine proteinases have been useful probes in the analysis of the molecular mechanisms of parasite invasion.[28,11] A surprising finding was obtained when we compared the anti-parasite effects of a hydrophobic peptidyl diazomethane inhibitor (Z-(SBz)Cys-Phe-CHN_2) with water soluble inhibitors, such as E-64, leupeptin, or cystatin C.[11] Unexpectedly, parasite invasion of HUVECs or CHO-B2R was only impaired by the membrane-permeable inhibitor.[11] The same results (unpublished data) were recently obtained with Mu-F-hf-VSPh, a vinylsulfone derivative that acts as a potent anti-parasite drug in the mouse model of infection.[49] Given the evidence that trypomastigotes are poorly endocytic[88] and that these flagellates accumulate cruzipain in the flagellar pocket,[9] the failure of hydrophilic inhibitors in preventing cellular invasion was interpreted as evidence that the kinin-releasing reaction most likely occurs in the enclosed areas formed by juxtaposition of host cell and parasite plasma membranes.[11] It is thus conceivable that the cruzipain molecules diffuse from the parasites' flagellar pocket into this intercellular space, during the penetration process. In this secluded microenvironment, the parasite protease may be possibly spared from physiological inactivation by soluble forms of plasma protease inhibitors (e.g., cystatins,[32,42] kininogens,[32,42] α_2-macroglobulin[72,73]). Moreover, active secretion by the parasites may continuously raise the concentration of cruzipain in the intercellular space. As the kinin peptide concentration builds up, the $[Ca^{2+}]_i$ response relayed through the bradykinin receptors is thought to stimulate lysosomal exocytosis,[89,89b] thus promoting the recruitment of these vesicles to the sites of parasite attachment.[90] We have suggested[11] that the discharge of the host lysosomal contents within these secluded intercellular spaces may generate a more acidic and reduced microenvironment, thus protecting the active site cysteine residues of cruzipain isoforms from oxidation and consequently increasing the half-life of the kinin-releasing enzyme.

Kinin-Receptors Mediate the Activation of Vascular Endothelium by Trypomastigotes

After decades of debate, studies in the late 90's have firmly demonstrated a relationship between persistence of myocardial parasitism and chronic chagasic myocardiopathy (CCM).[91,91b] Overlooked for many years, the proposition that microvascular lesions may be critically involved in CCM[92,93] has recently gained credence by three-dimensional confocal microscopy studies showing extensive microcirculatory abnormalities in myocardium autopsies from CCM patients.[94] These authors reported the presence of diffuse arteriolar dilatation and microvessel tortuosity in the myocardium of these patients, presumably resulting from lesions caused by impaired blood flow distribution.[94] Studies of the acute myocarditis in mice have previously shown that ischemic injury due to platelet aggregation and obstruction of capillaries could be linked to parasite activity.[95,96] In acutely infected dogs, fibrin microthrombi and endothelial cell lesions were also described at sites of contact with leukocytes in the myocardium, suggesting that the microangiopathy may be a consequence of parasite-induced immunopathology.[97]

Although rarely detected in tissue sections, infected cultures of endothelial cells release endothelin 1,[98] a potent vasoconstrictor that may be involved in the induction of vasospasm and myocardial ischemia. Furthermore, infected endothelial cells also secrete pro-inflammatory cytokines IL-1β and IL-6.[99] By activating the NFk-B pathway, these cytokines may upregulate the expression of vascular adhesion molecules on the endothelial lining,[100] thus ultimately driving the recruitment of leukocytes to sites of injury.[101,102]

At the present time, there is little knowledge about the host susceptibility factors underlying CCM. Studies in the canine model of indeterminate Chagas disease indicated that the myocarditis is rather mild and self-limited, and does not show obvious signs of microvascular pathology.[103] These authors suggested that progression towards CCM may result from sudden exacerbation of the chronic inflammatory process, possibly caused by abnormal shifts in the immunoregulatory balance in susceptible patients,[103,103b] rather than being characterized by a continuous process.

The description of a molecular pathway linking the activity of the kinin receptors to host cell susceptibility to *T. cruzi* infection raised questions as to their importance to the pathogenesis of CCM. In experimental models of ischemia/reperfusion,[105] there are solid data indicating that bradykinin exerts a cardioprotective role in the cardiovascular system[104] by signalling the constitutive kinin-receptors (B2). Thus, it is possible that ischemic lesions produced in the myocardium during the indeterminate stage of the disease may be more efficiently repaired due to low grade, albeit sustained, kinin production in inflamed tissues exposed to extracellular trypomastigotes. Conversely, activation responses conveyed through the kinin-receptors may render tissues increasingly susceptible to infection, thus ensuring mutual benefits to the host-parasite relationship (Fig. 7).

Although limited to animal models,[113] we have recently demonstrated that host peptidases critically modulate the parasite's ability to trigger pro-inflammatory responses via kinin receptors. In principle, a transitory drop in ACE or conversely, an increase in kininase I activity (Fig. 7), may subtle alter the homeostasis in inflamed tissues, perhaps allowing for a differential outgrowth of the *T. cruzi* clones that have the ability to activate the kinin cascade.[11,113] Under these circumstances, the parasite-host equilibrium may only be restored at the expense of a robust activation of cellular immune responses, a process that is inevitably linked to inflammation. Although speculative, it is possible that host susceptibility to *T. cruzi* infection may vary between chagasic patients due to genetically-determined fluctuations in the levels of the kinin degrading peptidases present in the cardiovascular tissues.[11,113]

Recently, we have started to investigate the outcome of the interaction of trypomastigotes with the vascular endothelium.[113] Using intravital microscopy as an experimental model, we first asked if purified cruzipain was able to trigger permeability increases on the post-capillary venules from the hamster cheek pouch.[106] Our early results indicated that the topical application of the protease did indeed induce plasma leakage. Importantly, the vascular response was blocked by pre-treatment of cruzipain with E-64, an irreversible inhibitor of papain-like proteases, thus demonstrating that the catalytic activity of the parasite protease was required.[106] Contrary to our initial expectation, plasma leakage was not inhibited when the tissues were treated with kinin-receptor antagonists. Instead, the vascular response was partially blocked by meparamine, a H1-histamine receptor blocker, suggesting that cruzipain may activate the microvasculature by additional pathways, perhaps leading to the triggering of mast cells.[106] More recently however, we have found out that cruzipain did not trigger the kinin cascade under these experimental conditions because kininogens, the substrate from which vasoactive kinins are released by proteolysis, do not accumulate in sufficient quantities in the "normal" (i.e., non-inflamed) cheek pouch tissues. Interestingly, however, kinin-receptor triggering accounts for the vigorous permeability increases that are observed when living trypomastigotes are added to the hamster cheek pouch (Svensjo et al, in preparation).[113] As opposed to assays performed with purified cruzipain, studies in wild type mice or in knockout animals with targeted deletion of kinin-receptors indicate that trypomastigotes indeed trigger microvascular responses through this pathway.[113] Of further interest, we also found that the vascular responses that the trypomastigotes induce in wild-type animals were blocked by treatment with Mu-F-hf-VSPh (Todorov et al, in preparation), a drug that has been previously shown to eradicate the parasites from infected mice.[48] Collectively, our recent studies suggest that trypomastigotes can promote the diffusion of blood-borne kininogens into extravascular sites, at early stages of the inflammatory reaction (Fig. 7). In addition, the effects of Mu-F-hf-VSPh support the concept that

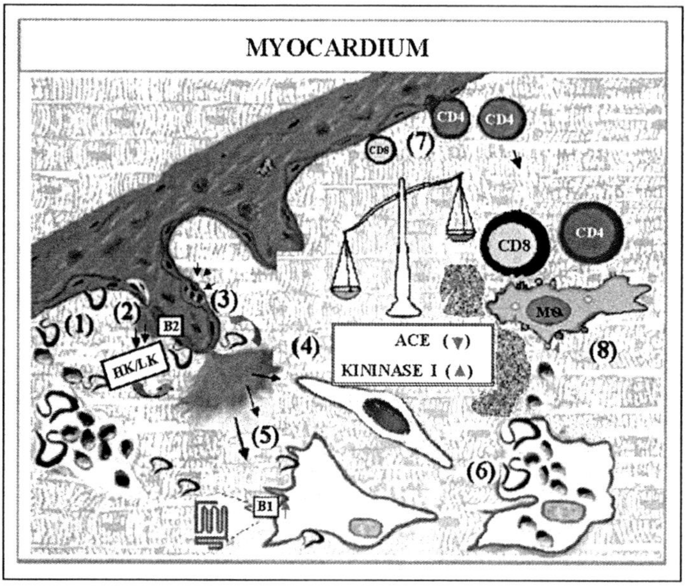

Figure 7. Modulation of cardiovascular pathology by kinin-degrading peptidases. The scheme illustrates a simplified view of the dynamics of the inflammatory response triggered by *T. cruzi*. This process is initiated when the infected host cells (e.g., cardiomyocytes) rupture (1), releasing trypomastigotes and amastigotes to the interstitial spaces. The trypomastigotes rapidly move away from the primary infection site (2) while the nonmotile amastigotes remain in the periphery of the original site of infection. A limited influx of plasma proteins occurs (2) as result of endothelium activation by chemokines and/or other pro-inflammatory mediators liberated by cruzipain[106] and/or other parasite molecules. The blood-borne kininogen molecules are retrieved in the extracellular matrix by heparan sulphate proteoglycans (2). The increased availability of kininogens in extravascular tissues that were exposed to kinin-releasing *T. cruzi* strains (clones) allows for increased infection of target cells expressing high levels of kinin-receptors.[1,113] Acting by the paracrine mode, the liberated kinin peptides bind with high-affinity to the constitutive B2 kinin-receptors expressed by cardiovascular cells[11,113] and/or immature dendritic cells[115] (not shown in the scheme). The action of kinin-degrading peptidases (e.g., Angiotensin Converting Enzyme, ACE or Neutral Endopeptidase, NEP) minimize the long range effects of these potent vasoactive peptides.[80,81] Temporal control of the kinin cascade is also achieved through B2 receptor down-regulation.[80,81] Increases of the kinin concentration at sites of parasite attachment stimulate vigorous $[Ca^{+2}]_i$ transients in the endothelial cells and the oedema is intensified. The parasites further exploit the increased availability of extravascular kininogen molecules, generating more kinins (3) through cruzipain.[11,113] Although counteracted by the immune response, parasite outgrowth in cardiovascular cells may be to some extent increased[11,113] Once infected, endothelial cells (3) may liberate endothelin 1,[98] a potent vasoconstrictor mediator, hence provoking ischemic lesions.[98,99] Given the indication that vasodilating kinin peptides exert a cardioprotective role in models of ischemia/reperfusion,[105,109] low-level stimulation of kinin-receptors may help the repair of myocardial lesions in infected tissues, perhaps contributing to the self-limited pathology which characterizes the indeterminate stage of Chagas′ disease.[103] In some patients, fluctuations in the levels of kinin-degrading enzymes in the myocardium may disturb this delicate host-parasite balance. The inflammation associated with new cycles of parasite outgrowth (4) leads to the upregulation of the B1 receptor subtype[113] by endothelial cells and other cell types, e.g., cardiac fibroblasts (5). In this setting, kininase I (carboxypeptidase M/N) assumes a pivotal role by converting the primary B2 agonists (bradykinin or kallidin) liberated by trypomastigotes into the agonists of the B1-kinin receptor (ie., [des-Arg] kinins).[113] Fibroblasts expressing B1 may respond to [des-Arg] kinin metabolites by inducing tissue fibrosis (*). Furthermore, the triggering of the inducible B1 receptor of endothelial cells (7) activates the NFk-B pathway and consequently promotes the upregulation of vascular adhesion molecules.[114] Leukocytes, e.g., lymphocytes from the CD8 and CD4 subsets, are then recruited to sites of infection(8), thus exacerbating inflammation.[113] (5). Since host kininase I is often upregulated in the inflamed tissues,[81,84] the kinin-releasing *T. cruzi* strains may engage the inducible B1 pathway to opportunistically invade non-phagocytic cells, including cardiomyocytes.[113] A robust T cell mediated response is therefore mobilized to restore the host-parasite balance (8), but this occurs at expense of excessive release of inflammatory cytokines. In some patients, the increased activity of ACE and/or other kinin-degrading peptidases may be increased to further protect the myocardium from being infected by kinin-releasing *T. cruzi* clones. Ischemic lesions, perhaps aggravated by the vasoconstriction caused by ACE and/or endothelin 1, may in the long term aggravate chronic cardiomyopathy.[94,98,99]

the pathogenic potential of *T. cruzi* trypomastigotes is critically influenced by the activity of cruzipain isoforms.

Concluding Remarks

The studies reviewed herein have suggested that the ability of *T. cruzi* to activate the kinin system is a tightly regulated process that depends on the interplay of multiple host and parasite factors.[110] The hypothesis predicts that the outgrowth of parasite clones that activate the kinin cascade may be enhanced due to the activation of the constitutively expressed B2 kinin receptor subtype, or alternatively, of the inducible B1 receptor subtype (Fig. 7). While engagement of the former may influence parasite tissue tropism at early stages of infection, the latter may be preferentially mobilized after the onset of tissue injury and inflammation.[113] Because the effects of the short-lived kinin peptides are modulated by multiple kinin-degrading peptidases, host susceptibility to the progressive forms of myocardial pathology may be, at least to some extent, linked to differences in the tissue levels and/or distribution of these host enzymes. Lenzi and co-workers[107] have previously drawn attention to the fact that the kidney parenchyma and lungs, i.e., highly vascularized tissues that abundantly express ACE,[108] are largely free from parasitism, while massive infection and inflammation develops in virtually every other organ. Thus, it is possible that in some tissues, ACE and other kinin-degrading peptidases may exert a host protective role, by reducing the infectivity of kinin-releasing *T. cruzi* clones. Considering that bradykinin acts as a cardioprotective molecule in models of ischemia/reperfusion,[105] the damage inflicted by potent vasoconstricting mediators, such as endothelin 1[98] may be counteracted by the low level production of the vasodilating kinin peptides in infected heart tissues. For the patients that, for unknown reasons, fail to sustain adequate levels of ACE or other kinin-degrading peptidases after years of infection, excessive stimulation of kinin-receptors may exacerbate cardiovascular pathology. For example, this may occur if the expression of host kininase I is enhanced in the inflamed myocardium (Fig. 7), because this peptidase can swiftly convert the released bradykinin or kallidin (i.e., agonists for B2) into an agonist (des-Arg bradykinin or des-Arg kallidin) for the B1 kinin-receptor subtype. Our recent studies[113] suggest that this strategy may allow the parasites to invade opportunistically macrophages, cardiac fibroblasts, vascular endothelial and smooth muscle cells, all of which upregulate the B1 kinin receptor subtype during inflammation. Moreover, triggering of the endothelial B1-kinin receptor by *T. cruzi* clones that have the competence to activate the kinin cascade may lead to the transcriptional activation of NFk-B.[111] Since this activation pathway upregulates the expression of vascular adhesion molecules,[114] blood-borne leukocytes may be recruited to sites of infection, thus further exacerbating inflammation. In order to compensate for the increased invasion of host cells that express the inducible B1 kinin-receptors, susceptible patients may rely on robust cellular immune responses, a process that is usually coupled to excessive release of NO and pro-inflammatory cytokines.[91,101,102] Immunopathology may thus converge with the microvascular lesions [92-94] triggered by the parasites, thus further amplifying tissue damage. As already mentioned, some chagasic patients may upregulate the expression of ACE and/or other kinin-degrading peptidases to protect heart cells from being so easily infected. However, this may in turn increase myocardium ischemia, which may in the long term aggravate the cardiomyopathy. Research on the factors that interfere with kinin homeostasis in the *T. cruzi* infected heart may offer new insights into the molecular pathogenesis of Chagas disease.

Acknowledgments

This research was sponsored by grants from PRONEX/MCT, PADCT, FAPERJ, CNPq, SRII-UFRJ and WHO-TDR (ID A10340). Important aspects of this research was carried out in collaboration with Dr. Ana Paula Lima, and with the crucial participation of several students from our lab. The long and productive partnership with Dr. Luiz Juliano (UNIFESP/EPM, São Paulo) is warmly acknowledged. I am also indebted to Dr. Werner Muller-Esterl (Frankfurt University, Germany) for his discussions and enthusiastic support to our project. I am grateful to Dr. Jim McKerrow (UCSF,USA), Dr. Jim Palmer (Axys Pharmaceutical, SF, USA), Dr. Francis Gauthier and Dr. G. Lalmanach (Tours University, France) for donation of various synthetic inhibitors and recombinant cruzain. Dr. Norma Andrews (Yale University) has kindly provided some of the prints obtained from ultrastructural studies.

References

1. Dorta ML, Ferreira AT, Oshiro ME et al. Ca^{2+} signal induced by Trypanosoma cruzi metacyclic trypomastigote surface molecules implicated in mammalian cell invasion. Mol Biochem Parasitol 1995; 73(1-2):285-9.
2. Burleigh BA, Andrews NW. A 120-kDa alkaline peptidase from Trypanosoma cruzi is involved in the generation of a novel Ca^{2+} -signaling factor for mammalian cells. J Biol Chem 1995; 270(10):5172-80.
3. Ruiz R, Favoreto S, Dorta ML, et al. Infectivity of Trypanosoma cruzi strains is associated with differential expression of surface glycoproteins with differential Ca^{2+} signaling activity. Biochem J 1998; 330:505-511.
4. Hall BS, Tam W, Sen R et al. Cell-specific activation of nuclear factor-kappa B by the parasite Trypanosoma cruzi promotes resistance to intracellular infection. Mol Biol Cell 2000; 11(1):153-60.
5. Ming M, Ewen ME, Pereira ME. Trypanosome invasion of mammalian cells requires activation of the TGF beta signaling pathway. Cell 1995; 82(2):287-96.
6. Tardieux I, Webster P, Ravesloot J et al. Lysosome recruitment and fusion are early events required for trypanosome invasion of mammalian cells. Cell 1992; 71(7):1117-30.
6b. Magdesian MH, Giordano R, Ulrich H et al. Infection by Trypanosoma cruzi. Identification of a parasite ligand and its host cell receptor. J Biol Chem 2001; 276(22):19382-9.
7. Moreno SN, Silva J, Vercesi AE et al. Cytosolic-free calcium elevation in Trypanosoma cruzi is required for cell invasion. J Exp Med 1994; 180(4):1535-40.
8. Burleigh BA, Caler EV, Webster P et al. A cytosolic serine endopeptidase from Trypanosoma cruzi is required for the generation of Ca^{2+} signaling in mammalian cells. J Cell Biol 1997; 136(3):609-20.
9. Murta AC, Persechini PM, Padron Td et al. Structural and functional identification of GP57/51 antigen of Trypanosoma cruzi as a cysteine proteinase. Mol Biochem Parasitol 1990; 43(1):27-38.
10. Del Nery E, Juliano MA, Lima AP et al. Kininogenase activity by the major cysteinyl proteinase (cruzipain) from Trypanosoma cruzi. J Biol Chem 1997; 272(41):25713-8.
10b. Lima, APCA, Almeida PC, Tersariol ILS et al. Heparan sulfate modulates kinin-release by Trypanosoma cruzi thorough the activity of cruzipain. J Biol Chem 2002; 277: 5875-81.
11. Scharfstein J, Schmitz V, Morandi V et al. Host cell invasion by Trypanosoma cruzi is potentiated by activation of bradykinin B(2) receptors. J Exp Med 2000; 192(9):1289-300.
12. Rawlings ND, Barrett AJ. Families of cysteine peptidases. Methods Enzymol 1994; 244:461-86.
13. Campetella O, Henriksson J, Aslund L et al. The major cysteine proteinase (cruzipain) from Trypanosoma cruzi is encoded by multiple polymorphic tandemly organized genes located on different chromosomes. Mol Biochem Parasitol 1992; 50(2):225-34.
14. Coombs GH, Mottram JC. Parasite proteinases and amino acid metabolism: possibilities for chemotherapeutic exploitation. Parasitology 1997; 114 Suppl:S61-80.
15. Aslund L, Henriksson J, Campetella O et al. The C-terminal extension of the major cysteine proteinase (cruzipain) from Trypanosoma cruzi. Mol Biochem Parasitol 1991; 45(2):345-7.
16. Eakin AE, Mills AA, Harth G et al. The sequence, organization, and expression of the major cysteine protease (cruzain) from Trypanosoma cruzi. J Biol Chem 1992; 267(11):7411-20.
17. Lima AP, Tessier DC, Thomas DY et al. Identification of new cysteine protease gene isoforms in Trypanosoma cruzi. Mol Biochem Parasitol 1994; 67(2):333-8.
17b. Duschak VG, Ciaccio M, Nassert JR et al. Enzymatic activity, protein expression, and gene sequence of cruzipain in virulent and attenuated Trypanosoma cruzi strains. J Parasitol 2001; 87(5):1016-22.

18. Cazzulo JJ, Martinez J, Parodi AJ et al. On the post-translational modifications at the C-terminal domain of the major cysteine proteinase (cruzipain) from Trypanosoma cruzi. FEMS Microbiol Lett 1992; 79(1-3):411-6.
19. Engel JC, Doyle PS, Palmer J et al. Cysteine protease inhibitors alter Golgi complex ultrastructure and function in Trypanosoma cruzi. J Cell Sci 1998; 111 (Pt 5):597-606.
20. Soares MJ, Souto-Padron T, De Souza W. Identification of a large pre-lysosomal compartment in the pathogenic protozoon Trypanosoma cruzi. J Cell Sci 1992 May;102 (Pt 1):157-67
21. McGrath ME, Eakin AE, Engel JC et al. The crystal structure of cruzain: a therapeutic target for Chagas disease. J Mol Biol 1995; 247(2):251-9.
22. McGrath ME. The lysosomal cysteine proteases. Annu Rev Biophys Biomol Struct 1999; 28:181-204.
23. Cazzulo JJ, Bravo M, Raimondi A et al. Hydrolysis of synthetic peptides by cruzipain, the major cysteine proteinase from Trypanosoma cruzi, provides evidence for self-processing and the possibility of more specific substrates for the enzyme. Cell Mol Biol (Noisy-le-grand) 1996; 42(5):691-6.
24. Lima AP, Scharfstein J, Storer AC et al. Temperature-dependent substrate inhibition of the cysteine proteinase (GP57/51) from Trypanosoma cruzi. Mol Biochem Parasitol 1992; 56(2):335-8.
24b. Salvati L, Mattu M, Polticelli F et al. Modulation of the catalytic activity of cruzipain, the major cysteine proteinase from Trypanosoma cruzi, by temperature and pH. Eur J of Biochem 2001; 268 (11):3253-58.
25. Gillmor SA, Craik CS, Fletterick RJ. Structural determinants of specificity in the cysteine protease cruzain. Protein Sci 1997; 6(8):1603-11.
26. Cazzulo JJ, Couso R, Raimondi A et al. Further characterization and partial amino acid sequence of a cysteine proteinase from Trypanosoma cruzi. Mol Biochem Parasitol 1989; 33(1):33-41.
27. Cazzulo JJ, Cazzulo Franke MC, Martinez J et al. Some kinetic properties of a cysteine proteinase (cruzipain) from Trypanosoma cruzi. Biochim Biophys Acta 1990; 1037(2):186-91.
27b. Stoka V, McKerrow JH, Cazzulo JJ et al. Substrate inhibition of cruzipain is not affected by the C-terminal domain. FEBS Lett 1998; 429:129- 33.
28. Meirelles MN, Juliano L, Carmona E et al. Inhibitors of the major cysteinyl proteinase (GP57/51) impair host cell invasion and arrest the intracellular development of Trypanosoma cruzi in vitro. Mol Biochem Parasitol 1992; 52(2):175-84.
29. Nery ED, Juliano MA, Meldal M et al. Characterization of the substrate specificity of the major cysteine protease (cruzipain) from Trypanosoma cruzi using a portion-mixing combinatorial library and fluorogenic peptides. Biochem J 1997; 323 (2):427-33.
30. Serveau C, Lalmanach G, Juliano MA et al. Investigation of the substrate specificity of cruzipain, the major cysteine proteinase of Trypanosoma cruzi, through the use of cystatin-derived substrates and inhibitors. Biochem J 1996; 313 (3):951-6.
31. Hellman U, Wernstedt C, Cazzulo JJ. Self-proteolysis of the cysteine proteinase, cruzipain, from Trypanosoma cruzi gives a major fragment corresponding to its carboxy-terminal domain. Mol Biochem Parasitol 1991; 44(1):15-21.
32. Stoka V, Nycander M, Lenarcic B et al. Inhibition of cruzipain, the major cysteine proteinase of the protozoan parasite, Trypanosoma cruzi, by proteinase inhibitors of the cystatin superfamily. FEBS Lett 1995; 370(1-2):101-4.
33. Itow S, Camargo EP. Proteolytic activites in cell extracts of Trypanosoma cruzi. J Protozool 1977; 24(4):591-5.
34. Rangel HA, Araujo PM, Camargo IJ et al. Detection of a proteinase common to epimastigote, trypomastigote and amastigote of different strains of Trypanosoma cruzi. Tropenmed Parasitol 1981; 32(2):87-92.
35. Mendonça-Previato L, Gorin PA, Braga AF et al. Chemical structure and antigenic aspects of complexes obtained from epimastigotes of Trypanosoma cruzi. Biochemistry 1983; 22(21):4980-7.
36. Scharfstein J, Rodrigues MM, Alves CA et al. Trypanosoma cruzi: description of a highly purified surface antigen defined by human antibodies. J Immunol 1983; 131(2):972-6.
37. Scharfstein J, Luquetti A, Murta AC et al. Chagas disease: serodiagnosis with purified Gp25 antigen. Am J Trop Med Hyg 1985; 34(6):1153-60.
38. Scharfstein J, Schechter M, Senna M et al. Trypanosoma cruzi: characterization and isolation of a 57/51,000 m.w. surface glycoprotein (GP57/51) expressed by epimastigotes and bloodstream trypomastigotes. J Immunol 1986; 137(4):1336-41.

39. Gazzinelli RT, Leme VM, Cancado JR et al. Identification and partial characterization of Trypanosoma cruzi antigens recognized by T cells and immune sera from patients with Chagas disease. Infect Immun 1990; 58(5):1437-44.
40. Metzner SI, Sousa MC, Hellman U et al. The use of UDP-Glc:glycoprotein glucosyltransferase for radiolabeling protein-linked high mannose-type oligosaccharides. Cell Mol Biol (Noisy-le-grand) 1996; 42(5):631-5.
41. Martinez J, Campetella O, Frasch AC et al.The reactivity of sera from chagasic patients against different fragments of cruzipain, the major cysteine proteinase from Trypanosoma cruzi, suggests the presence of defined antigenic and catalytic domains. Immunol Lett 1993; 35(2):191.
41b. Giordanengo L, Maldonado C, Rivarola HW et al. Induction of antibodies reactive to cardiac myosin and development of heart alterations in cruzipain-immunized mice and their offspring. Eur J Immunol 2000; 30(11):3181-9.
42. Scharfstein J, Abrahamson M, de Souza CB et al. Antigenicity of cystatin-binding proteins from parasitic protozoan. Detection by a proteinase inhibitor based capture immunoassay (PINC-ELISA). J Immunol Methods 1995; 182(1):63-72.
43. Arnholdt, AC, Scharfstein, J. Immunogenicity of Trypanosoma cruzi cysteine proteinase. Rs Immunology 1991; 142(2);146-51.
44. Cazzulo JJ, Frasch AC. SAPA/trans-sialidase and cruzipain: two antigens from Trypanosoma cruzi contain immunodominant but enzymatically inactive domains. FASEB J 1992; 6(14):3259-64.
45. Arnholdt AC, Piuvezam MR, Russo DM et al. Analysis and partial epitope mapping of human T cell responses to Trypanosoma cruzi cysteinyl proteinase. J Immunal 1993; 151(6):3171-9.
46. Harth G, Andrews N, Mills AA et al. Peptide-fluoromethyl ketones arrest intracellular replication and intercellular transmission of Trypanosoma cruzi. Mol Biochem Parasitol 1993; 58(1):17-24.
47. Franke de Cazzulo BM, Martinez J, North MJ et al. Effects of proteinase inhibitors on the growth and differentiation of Trypanosoma cruzi. FEMS Microbiol Lett 1994; 124(1):81-6.
48. McKerrow JH, McGrath ME, Engel JC. The cysteine protease of Trypanosoma cruzi as a model for antiparasite drug design. Parasitol Today 1995; 11:279-282.
49. Engel JC, Doyle PS, Hsieh I et al. Cysteine protease inhibitors cure an experimental Trypanosoma cruzi infection. J Exp Med 1998; 188(4):725-34.
50. Tomas AM, Kelly JM. Stage-regulated expression of cruzipain, the major cysteine protease of Trypanosoma cruzi is independent of the level of RNA. 1996; 76(1-2):91-103.
51. Lima APCA , Reis FCG, Serveau C et al. Cysteine protease isoforms from Trypanosoma cruzi, cruzipain 2 and cruzain, present different substrate preference and susceptibility to inhibitors. Mol Biochem Parasitol 2001; 114(1):41-52.
52. Engel JC, Torres C, Hsieh I, Doyle PS et al. Upregulation of the secretory pathway in cysteine protease inhibitor-resistant Trypanosoma cruzi. J Cell Sci 2000; 113(8):1345-54.
53. Yong V, Schmitz V, Vannier-Santo MA et al. Altered expression of cruzipain and a cathepsin B-like target in a Trypanosoma cruzi cell line displaying resistance to synthetic inhibitors of cysteine-proteinases. Mol Biochem Parasitol 2000; 09:47-59.
54. Cazzulo JJ, Stoka V, Turk V. Cruzipain, the major cysteine proteinase from the protozoan parasite Trypanosoma cruzi. Biol Chem 1997; 378:1-10.
55. Parodi AJ, Labriola C, Cazzulo JJ. The presence of complex-type oligosaccharides at the C-terminal domain glycosylation site of some molecules of cruzipain. Mol. Biochem Parasitol 1995; 69(2):247-55.
56. Martinez J, Henriksson J, Ridaker M et al. Polymorphisms of the genes encoding cruzipain, the major cysteine proteinase of Trypanosoma cruzi, in the region encoding the C-terminal domain. FEMS Microbiol Lett 1998; 159(1):35-9.
57. Souto-Padron T, Campetella OE, Cazzulo JJ et al. Cysteine proteinase in Trypanosoma cruzi: immunocytochemical localization and involvement in parasite-host cell interaction. J Cell Sci 1990; 96(Pt 3):485-90.
58. Fresno M, Hernandez-Munain C, de-Diego J et al. Trypanosoma cruzi: identification of a membrane cysteine proteinase linked through a GPI anchor. Braz J Med Biol Res 1994; 27(2):431-7.
59. Parussini F, Duschak VG, Cazzulo JJ. Membrane-bound cysteine proteinase isoforms in different developmental stages of Trypanosoma cruzi. Cell Mol Biol (Noisy-le-grand) 1998; 44(3):513-9.

60. Campetella O, Martinez J, Cazzulo JJ. A major cysteine proteinase is developmentally regulated in Trypanosoma cruzi. FEMS Microbiol Lett 1990; 55(1-2):145-9.
61. Tomas AM, Miles MA, Kelly JM. Overexpression of cruzipain, the major cysteine proteinase of Trypanosoma cruzi, is associated with enhanced metacyclogenesis. Eur J Biochem 1997; 244(2):596-603.
62. Monteiro ACS, Abrahamson M., Lima APCA et al. Identification, characterization and localization of chagasin, a tight-binding cysteine proteinase inhibitor in Trypanosoma cruzi. J Cell Sci 2001; 114 (21) : 3933-42.
63. Labriola C, Sousa M, Cazzulo JJ. Purification of the major cysteine proteinase (cruzipain) from Trypanosoma cruzi by affinity chromatography. Biol Res 1993; 26 :101-7l.
64. Nobrega OT, Santos Silva MA, Teixeira AR et al. Cloning and sequencing of tccb, a gene encoding a Trypanosoma cruzi cathepsin B-like protease. Mol Biochem Parasitol 1998; 97(1-2):235-40.
65. Huete-Perez JA, Engel JC, Brinen LS et al. Protease trafficking in two primitive eukaryotes is mediated by a prodomain protein motif. J Biol Chem 1999; 274(23): 16249-56.
66. Brooks DR, Tetley L, Coombs GH et al. Processing and trafficking of cysteine proteases in Leishmania mexicana. J Cell Sci 2000; 113(Pt 22):4035-41.
66b. Nascimento AE, de Souza W. High resolution localization of cruzipain and Ssp4 in Trypanosoma cruzi by replica staining label fracture. Biol Cell 1996;86(1):53-8.
67. Bonaldo MC, Scharfstein J, Murta AC et al. Further characterization of Trypanosoma cruzi GP57/51 as the major antigen expressed by differentiating epimastigotes. Parasitol Res 1991; 77(7):567-71.
68. Venturini G, Salvati L, Muolo M et al. Nitric oxide inhibits cruzipain, the major papain-like cysteine proteinase from Trypanosoma cruzi. Biochem Biophys Res Commun 2000; 270(2):437-41.
69. Stoka V V, Turk B, Schendel SL et al. Lysosomal protease pathways to apoptosis: cleavage of bid, not Pro-caspases, is the most likely route. J Biol Chem 2000; 276(5):3149-57.
70. Freire-de-Lima, CG, Nascimento, DO, Soares, MB et al. Uptake of apoptotic cells drives the growth of pathogenic trypanosome in macrophages. Nature 2000; 403 (6766):194-203.
71. Scharfstein J, Morrot, A. A role for extracellular amastigotes in the Immunopathology of Chagas disease. Mem Inst Osw Cruz 1999; 94 (suppl 1):51-63.
72. Morrot A, Strickland DK, Higuchi MdL et al. Human T cell responses against the major cysteine proteinase (cruzipain) of Trypanosoma cruzi: role of the multifunctional alpha2-macroglobulin receptor in antigen presentation by monocytes. Int Immunol 1997; 9(6):825-34.
72b. Basu S, Binder RJ, Ramalingam T, et al. CD91 is a common receptor for heat shock proteins gp96, hsp90, hsp70, and calreticulin. Immunity 2001;14:303-13.
73. Coutinho CMLM,Cavalcanti, GH, van Leuven F. et al. Alpha-2-macroglobulin binds to the surface of Trypanosoma cruzi. Parasitol Res 1999; 83:144-50.
74. Araujo-Jorge TC, Sampaio EP, and de Souza, W. Trypanosoma cruzi: inhibition of host cell uptake of infective bloodtream forms by alpha-2-macroglobulin. Z Paratenkd 1986; 72:323-29.
75. Coutinho CMLM, van Leuven F , and Araujo-Jorge, TC. Detection of alpha-2-macroglobulin in the heart of mice infected with Trypanosoma cruzi. Parasitol Res 1999; 85:249-55.
76. Waghabi MC, Coutinho CM, Soeiro MN et al. Incresaed Trypanosoma cruzi invasion and heart fibrosis associated with high transforming growth factor beta levels in mice deficient in alpha(2)-macroglobulin. Infect Immun 2002; 70:5115-23.
76b. Schnapp AR, Eickhoff CS, Scharfstein J et al. Induction of B- and T-cell responses to cruzipain in the murine model of Trypanosoma cruzi infection. Microbes Infect 2002; 4:805-13.
76c. Hoft DF, Eickhoff CS. Type 1 immunity provides optimal protection against both mucosal and systemic Trypanosoma cruzi challenges. Infect Immun 2002; 70:6715-25.
76d. Giordanengo L, Guinazu N, Stempin C et al. Cruzipain, a major Trypanosoma cruzi antigen, conditions the host immune response in favor of parasite. Eur J Immunol 2002; 32:1003-11.
77. Scharfstein J, Mendonça-Previato L, Borojevic R. Human antibody response to a surface glycoprotein isolated from Trypanosoma cruzi. Proc Pontificia Academia Scientiarvm 1982; 51:150-4.
78. Piras MM, Piras R, Henriquez D et al. Changes in morphology and infectivity of cell culture-derived trypomastigotes of Trypanosoma cruzi. Mol Biochem Parasitol 1982; 6(2):67-8.
79. Quitterer U, Schroeder C, Müller-Esterl W et al. Effects of bradykinin and endothelin-1 on the calcium homeostasis of mammalian cells. J Biol Chem 1995; 270:1992-9.

80. Bhoola KD, Figueroa CD and Worthy K. Bioregulation of kinins: kallikreins, kininogens, and kininases. Pharmacol Rev 1992; 44:1-80.
81. Couture R. and Lindsey CJ. Brain kallikrein-kinin system: from receptors to neuronal pathways and physiological functions. Quirion R, Blorklund A, Hokfelt T, eds. Handbook of Chemical Neuroanatomy Peptide Receptors. Part 1; Vol. 16. Elsevier Science, BV, 2000: 241-300.
82. Schmaier A. Plasma Kallikrein/kinin system: a revised hypothesis for its activation and its physiological contributions. Current Opinion on Hematology 2000; 7:261-5.
83. Kozik A, Moore RB, Potempa J et al. A novel mechanism for bradykinin production at inflammatory sites. J Biol Chem 1998; 273:33224-9.
84. Marceau F. Kinin B1 receptors: a review. Immunopharmacol 1995; 30:1-26.
85. Kaufmann J, Haasemann M, Modrow S et al. Structural dissection of the multi-domain kininogens. Fine mapping of the target epitopes of antibodies interfering with their functional properties. J Biol Chem. 1993; 268:9079-91.
86. Herwald H, Hasan AAK, Godovac-Zimmermann J et al. Identification of an endothelial cell binding site on kininogens' domain. J Biol Chem 1995; 270:14634-42.
87. Renné T, Dedio J, David G et al. H-Kininogen utilizes heparan sulfate proteoglycans for accumulation on Endothelial Cells. J Biol Chem 2000; 275:33688- 96.
87b. Renne T , Muller-Esterl W. Cell surface-associated chondroitin sulfate proteoglycans bind contact phase factor H-kininogen. FEBS Letters 2001; 500:36-40.
88. de Souza W. Structural organization of the cell surface of pathogenic protozoa. Micron 1995; 26:405-30.
89. Andrews NW. Regulated secretion of conventional lysosomes. Trends in Cell Biology 2000; 10:316-21.
89b. Burleigh BA, Woolsey AM. Cell signalling and Trypanosoma cruzi invasion. Cell Microbiol 2002; 4:701-11.
90. Rodriguez A, Samoff E, Rioult MG et al. Host cell invasion by trypanosomes requires lysosomes and microtubule/kinesin-mediated transport. J Cell Biol 1996; 134:349-62.
91. Tarleton RL, Zhang L, Downs MO. "Autoimmune rejection" of neonatal heart transplants in experimental Chagas disease is a parasite-specific response to infected host tissue. Proc Natl Acad Sci USA 1997; 94:3932-37.
91b. Palomino SA, Aiello VD, Higuchi ML. Systematic mapping of hearts from chronic chagasic patients: the association between the occurrence of histopathological lesions and Trypanosoma cruzi antigens. Ann Trop Med Parasitol 2000; 94:571-79.
92. Rossi M. Microvascular changes as a cause of chronic cardiomyopathy in Chagas disease. Am Heart Journal 1990; 120:233-36.
93. Morris SA, Tanowitz H, Wittner M et al. Pathophysiological insights into the cardiomyopathy of Chagas disease. Circulation 1990; 82:1900-09.
94. Higuchi ML, Fukasawa S, De Brito T et al. Different microcirculatory and interstitial matrix patterns in idiopathic dilated cardiomyopathy and Chagas disease: a three dimensional confocal microscopy study. Heart 1999; 81:0-6.
95. Rossi MA, Gonçalves S, Ribeiro dos Santos R. Experimental Trypanosoma cruzi cardiomyopathy in BALB/C mice: the potential role of intravascular platelet aggregation in its genesis. Am J Pathol 1984; 114:209-16.
96. Factor SM,Cho S, Wittner M et al. Abnormalities of the coronary microcirculation in acute murine Chagas disease. Am J Trop Med Hyg 1985; 34:246-53.
97. Andrade ZA, Andrade SG, Correa R et al. Myocardial changes in acute Trypanosoma cruzi infection: ultrastructural evidence of immune damage and the role of microangiopathy. Amer J Pathol 1994; 144:1403-11.
98. Wittner M, Christ GJ, Huang L et al. Trypanosoma cruzi induces endothelin release from endothelial cells. J Infect Dis 1995; 171:493-97.
99. Tanowitz HB, Gumprecht JP, Spurr D et al. Cytokine gene expression of endothelial cells infected with Trypanosoma cruzi. J Infect Dis 1992; 166:598-603.
100. Huang H, Calderon,TM, Berman, JW et al. Infection of endothelial cells with Trypanosoma cruzi activates NF-κB and induces vascular adhesion molecule expression. Infect Immunity 1999; 67:5434-40

101. Zhang L, Tarleton RL. Persistent production of inflammatory and anti- inflammatory cytokines and associated MHC and adhesion molecule expression at the site of infection and disease in experimental Trypanosoma cruzi infections. Exp. Parasitol 1996; 84: 203-13.
102. Sunnemark D, Frostegard J, Orn A et al. Cellular and cytokine characterization of vascular inflammation in CBA/J mice chronically infected with Trypanosoma cruzi. Scand J Immunol 1998; 48:480-4.
103. Andrade ZA, Andrade SG, Sadigursky M et al. The indeterminate phase of Chagas disease: ultrastructural characterization of cardiac changes in the canine model. Am J Trop Med Hyg 1997; 57:328-36.
103b. Correa-Oliveira R, Gomes J, Lemos EM et al. The role of the immune response on the development of severe clinical forms of human Chagas disease. Mem Inst Oswaldo Cruz 1999; 94(Suppl 1):253-5.
104. Figueroa CD, Marchant A, Novoa et al. Differential distribution of bradykinin B2 receptors in the rat and human cardiovascular system. Hypertension 2001; 37(1):110-20.
105. Yang XP, Liu YH, Mehta D et al. Diminished Cardioprotective Response to Inhibition of Angiotensin-Converting Enzyme and Angiotensin II Type 1 Receptor in B2 Kinin Receptor Gene Knockout Mice. Circ Res 2001; 88:1072-79.
106. Svensjo E, Cyrino FZ, Juliano L et al. Plasma leakage induced in postcapillary venules by the major cysteine-proteinase from Trypanosoma cruzi and its modulation by H1-blocker mepyramine. Microvasc Res 1997; 54(1):93-7.
107. Lenzi, HL, Oliveira DN, Lima MT et al.. Trypanosoma cruzi: paninfectivity of CL strain during murine acute infection. Experim Parasitol 1996; 84:16-27.
108. Danilov SM, Faerman AI, Printseva OY et al. Immunohistochemical study of angiotensin-converting enzyme in human tissues using monoclonal antibodies. Histochemistry 1987; 87:487-90.
109. Tschope CS, Heringer-Walther S, Walther TS. Regulation of the kinin-receptors after induction of myocardial infarction: a mini-review. Braz J Med Biol Res 2000; 33:701-8.
110. de Diego JA, Palau MT, Gamallo C et al. Relationship between histopathological findings and phylogenetic divergence in Trypanosoma cruzi. Trop Med Int Health 1998; 3:222-33.
111. Xie P, Browning D, Hay N et al. Activation of NFk-B by bradykinin through a Gaq- and Gbg-dependent pathway that involves phosphoinositide 3-kinase and Akt. J Biol Chem 2000; 275:24907-14
112. Lalmanach G, Mayer R, Serveau C et al. Biotin-labelled diazomethylketones derived from the substrate-like sequence of cystatin: a versatile tool to target the active site of cruzipain, the main cysteine proteinase of Trypanosoma cruzi. Biochem J 1996. 318:395-399.
113. Todorov AG, Andrade D, Pesquero JB et al. Trypanosoma cruzi induces edematogenic responses in mice and invades cardiomyocytes and endothelial cells in vitro by activating distinct kinin receptor (B1/B2) subtypes. FASEB J 2003; 17:73-5.
114. McLean P, Ahluvalia A and Perretti, A. Association between kinin B1 receptor expression and leukocyte trafficking across mesenteric postcapillary venules. J Exp Med 2001; 192:367-80.
115. Aliberti J, Viola JPB, Abreu ADV et al. Cutting edge: bradykinin induces IL-12 production by dendritic cells: a "danger" signal that drives Th1 polarization. J Immunol 2003; in press.

CHAPTER 9

Trypanosoma cruzi trans-Sialidase:

A Cytokine Mimetic (Parasitokine)

Wenda Gao and Miercio A. Pereira

Abstract

Cytokines are small soluble proteins with high potency in orchestrating host immune responses during stress, injury, tumorgenesis, and infection. Studies in animal models and humans have generated a large body of evidence correlating specific anti-parasite immune responses with cytokine patterns. For instance, control of intracellular pathogens requires cellular immune responses mediated by type 1 cytokines (IL-2, IFN-γ and others) while control of extracellular pathogens necessitates humoral immune responses accompanied by preferential type 2 cytokines (IL-4 and others). Cytokine profiles arise as a consequence of the antigenic nature of parasite molecules exposed to the host immune system. However, the harmony of cytokine secretion can be sabotaged by the parasites through molecular mimicry. It is well known that some viruses encode soluble molecules structurally and functionally similar to anti-inflammatory cytokines and pro-inflammatory cytokine receptors, named virokines and viroceptors, respectively. A novel concept that protozoan and worm parasites can produce molecules structurally and/or functionally akin to host cytokines, named parasitokines, is emerging. We describe here four model parasitokines derived from three distinct parasites: 1) *Trypanosoma cruzi: trans*-sialidase (TS), an inducer of cytokine secretion in normal endothelial cells and immune cells, of immunoglobulin production in normal B lymphocytes, and of T lymphocyte activation through antigen-presenting cells. 2) *T. cruzi*: proline racemase, a polyclonal activator of B lymphocytes. 3) *Trypanosoma brucei*: T lymphocyte triggering factor (TLTF), a polyclonal activator of $CD8^+$ T lymphocytes and an inducer of IFN-γ, a cytokine that promotes growth of *T. brucei.* And 4) *Brugia malayi*: Bm-MIF, a 12.5 kDa peptide homologue of macrophage migration inhibitory factor (MIF). Parasitokines, like conventional cytokines, could be important promoters of parasitism. Indeed, experiments in vivo showed that TS can enhance virulence of *T. cruzi* and of transgenic *Leishmania major*.

Overview of Host-Pathogen Interaction

Mammalian hosts are constantly exposed to a variety of potentially harmful microbes. To control infection, hosts activate the innate and adaptive arms of the immune system and utilize both cellular and humoral resources in appropriately polarized immune responses. Much attention has been paid to the immune regulation by the host during infection. However, the host-pathogen interaction is a highly dynamic process with both players affecting each other at

Molecular Mechanisms in the Pathogenesis of Chagas Disease, edited by John M. Kelly.

multiple levels. Infectious agents have developed various strategies to counter potent immune attacks from the host. These strategies fall into two major categories, summarized in Table I.

First, pathogens can reduce or alter their own immunogenicity through mechanisms such as molecular mimicry[1,2] and antigenic variation.[3,4] Second, pathogens can actively inhibit host immune responses in ways that are more aggressive than passive immune evasion. For instance, some gram-negative bacteria (*Salmonella*, *Shigella* and *Yersinia*) trigger their unique type III secretion systems upon contact with host cells to directly deliver virulence proteins into the host cell cytosol.[5-7] Some of these effector proteins are protein kinases and phosphatases,[8] which allow a pathogen to directly interfere with host cell signaling pathways.[9,10] The so-called effector protein Yops (*Yersinia* outer proteins), when injected into host cell cytosol by the bacteria, can disrupt cellular immune defense functions such as NF-κB activation, TNF-α release, and oxidative burst generation.[10] Furthermore, Yops can induce a great percentage of host phagocytes to undergo apoptosis, thereby allowing *Yersinia* to resist phagocytosis and grow extracellularly.[11-13]

Even more striking, pathogens can secrete/shed molecules mimicking the functions of host cytokines and chemokines to sabotage host-specified anti-parasite immune regulation. This article will focus on the latter form of host immunity subversion, in particular on the biological activities of the parasitokine TS encoded by *Trypanosoma cruzi*.

Important Roles of Cytokines in Host Defense

Evidence from animal models and human studies indicates that many soluble proteins secreted from various cell types orchestrate the highly regulated innate and adaptive immune responses. These proteins, now called cytokines and chemokines, are small molecules (~10 to 30 kDa) with high potency in altering the behavior and properties of the cells that produce them or of other cells. They are indispensable for the proper functioning of the immune system.

One important role of cytokines is to interconnect innate and adaptive immune responses. For instance, early growth of the intracellular protozoan *Toxoplasma gondii* in immunocompetent hosts is controlled by interferon-γ (IFN-γ) produced by IL-12-stimulated natural killer (NK) cells.[14,15] In contrast, later in infection, CD8$^+$ T lymphocytes appear to be the major effectors of adaptive resistance, but their protective function is also mediated by IFN-γ.[16,17] In general, cytokines such as IL-1, IL-6 and TNF-α are produced within hours by various types of cells in response to many infectious agents. These cytokines promote inflammation and acute phase responses and also play important roles in the adaptive immunity.[18,19] When mice are deficient in these cytokines or their receptors, they become highly susceptible to infections produced by various pathogens.[20-23]

Modulation of Host Immune Regulation by Cytokine Analogues

Given the important roles of cytokines as modulators of host defense, it is not unexpected that pathogens attempt to undermine host immunity by mimicking the actions of relevant cytokines and cytokine receptors of the host. Virokines and bacteriokines, reviewed extensively elsewhere (refs. 24-26), are briefly summarized below, while parasite mimics of mammalian cytokines and growth factors (parasitokines) are reviewed for the first time here. One important feature of pathogen-derived cytokine mimics is that they affect immune functions independent of being presented on MHC to provoke antigenic recognition by specific lymphocytes. Hence, cytokine mimics can target innate immune responses before adaptive immune responses take place.

Virokine and Viroceptors

Virokines include cytokine homologues of IL-6, IL-10 and IL-17,[27-32] while viroceptors include receptor homologues for TNF,[33-37] IFN-α/γ,[38-40] IL-1β[36,41] and IL-18.[42,43] Sequence

Table 1. Some microbial strategies to undermine host defense

Categories	Strategies	Consequences	Examples
Pathogens reduce or alter their own antigenicity	Latency	Long term presence of pathogen without eliciting immune response	Human herpes viruses (HSV-1, EBV)
(Passive evasion)	Molecular mimicry	Host recognition of certain microbial epitopes as self	Microbial epitopes that share cross-reactivity with those of host molecules
	Antigenic variation	Transient and limited efficacy of neutralizing antibodies against a specific microbial epitope	Variable surface glycoprotein (VSG) of African trypanosomes
Pathogens resist host immune effector function	Delivery of toxins or signaling molecules into host cell cytosol	Loss of effector functions or apoptosis of host immune cells	Type III secretion system of some gram-negative bacteria
(Active subversion)	Disruption of cytokine/chemokine signaling or inhibition of complement activation	Blockade of cytokine signaling; inactivation of complement; immune deviation; polyclonal lymphocyte activation	Virokine/viroceptor; Bacteriokine; Parasitokine

homology studies suggest that virokines and viroceptors were acquired from the host genome during evolution. Virokines and viroceptors appear to provide the viruses with advantages to evade host immunity, thereby contributing to the pathogenesis of viral infection.[44,45]

Bacteriokine

Many bacterial pathogens modulate host cytokine synthesis as a major virulence mechanism. It was believed that this capacity is due to a small number of bacterial cell wall components, such as lipopolysaccharide (LPS). However, it is now clear that bacteria contain a large number of diverse molecules, named bacteriokines, that can selectively induce the synthesis of both pro-inflammatory and immunomodulatory/anti-inflammatory cytokines.[25,46] Apart from the cytokine-inducing components associated with the bacterial cell wall, such as lipoproteins, carbohydrates, and lipids, several bacterial exotoxins are well-studied superantigens for host T cells.[47] These superantigens potently induce secretion of a diverse array of cytokines from T cells carrying specific Vβ T cell receptors (TCR). In vivo, such a storm of cytokine release upon interaction of T cells with superantigens can cause lethal shock of an animal.[48] Moreover, T cells activated by superantigens become anergic,[49,50] or even undergo activation-induced apoptosis.[51]

Parasitokine

Four parasite proteins fit the definition of parasitokine, or parasite-specified mimics of host cytokines and growth factors: 1) *T. cruzi trans*-sialidase (TS), 2) *T. cruzi* proline racemase,

3) *T. brucei* T lymphocyte triggering factor (TLTF), and 4) *Brugia malayi* Bm-MIF, a homologue of macrophage migration inhibitory factor (MIF).

T. cruzi trans-Sialidase (TS)

Carbohydrate Specificity and Primary Structure of TS

T. cruzi expresses a developmentally regulated *trans*-neuraminidase (sialidase) (TS) present at highest concentration on invasive trypomastigotes.[52,53] TS binds to donor substrates bearing α-2,3-linked sialic acid and transfers the carbohydrate to acceptor substrates with terminal nonreducing β-Gal residues.[54,55] TS consists of an N-terminal domain containing the catalytic site and a C-terminal domain consisting of long tandem repeats (LTR). Each tandem repeat has a highly conserved sequence of 12 amino acids (DSSAH$^{S}/_{G}$TPSTP$^{V}/_{A}$)[56] (Fig. 1). The catalytic domain of TS harbors a fibronectin type III unit generally associated with protein-protein interactions.[56] TS is attached to the parasite plasma membrane via a GPI anchor, and hence it is readily shed into the surrounding milieu by host and parasite phospholipases.[52,53]

TS As an Indirect Mediator of Infection: Virulence-Enhancement in Mouse Models of Chagas Disease and Cutaneous Leishmaniasis

Several lines of evidence suggest that surface-bound TS promotes attachment of *T. cruzi* to mammalian cells as a prelude to invasion.[57-59] Attachment is achieved by the ability of TS to bind sialic acid in a lectin-like manner.[57]

However, TS can also promote parasitism by mechanisms indirectly related to *T. cruzi*-host cell interactions. Administration of tiny amounts of the enzyme (10-20 ng/mouse) into the footpads of normal mice rendered the animals highly susceptible to *T. cruzi* invasion.[60] TS-enhanced virulence cannot be explained at the level of parasite adhesion to host cells because the neuraminidase enhanced virulence when injected into a footpad distinct from the one inoculated with parasites.[60] TS-enhanced virulence seemed to be mediated by host lymphocytes because its enhancement was not pronounced in *Scid* or nude mice. Moreover, the optimal virulence enhancement occurred when TS was injected 1-2 hours prior to, but not concurrent or after, infection. These results are consistent with the idea that TS can enhance *T. cruzi* virulence through immunological events occurring before the onset of adaptive responses.[60]

Confirmation of the virulence-enhancing activity of TS was provided by the heterologous expression of the enzyme in another protozoan parasite, *Leishmania major.* Ectopic expression of TS in *L. major* greatly increased virulence in a murine model of cutaneous leishmaniasis.[61] In this model, TS did not seem to alter *Leishmania* attachment to, and replication in, host macrophages.[61] Rather, the results with transgenic *Leishmania* are consistent with TS enhancing virulence by deregulating immune cell functions by mimicking host cytokines.

TS Is a Cytokine Secretagogue

TS can induce secretion of the cytokine IL-6 in human endothelial cells and peripheral blood mononuclear cells (PBMC) from healthy individuals.[62] In addition, TS can also stimulate IL-6 secretion in splenocytes and bone marrow cells from naïve mice as well as in a mouse macrophage cell line (W. Gao, H. Wortis and M. Pereira, manuscript in preparation). Such IL-6 secretagogue activity is independent of the *trans*-sialidase activity as it resides in the noncatalytic C-terminal long tandem repeat (LTR) domain of the enzyme.[62]

The cytokine secretagogue action of TS resembles the activity of authentic cytokines, which usually stimulate secretion of other cytokines in various cell types, as is the case of IL-1, which stimulates secretion of IL-6 from fibroblasts and other cell types.[63]

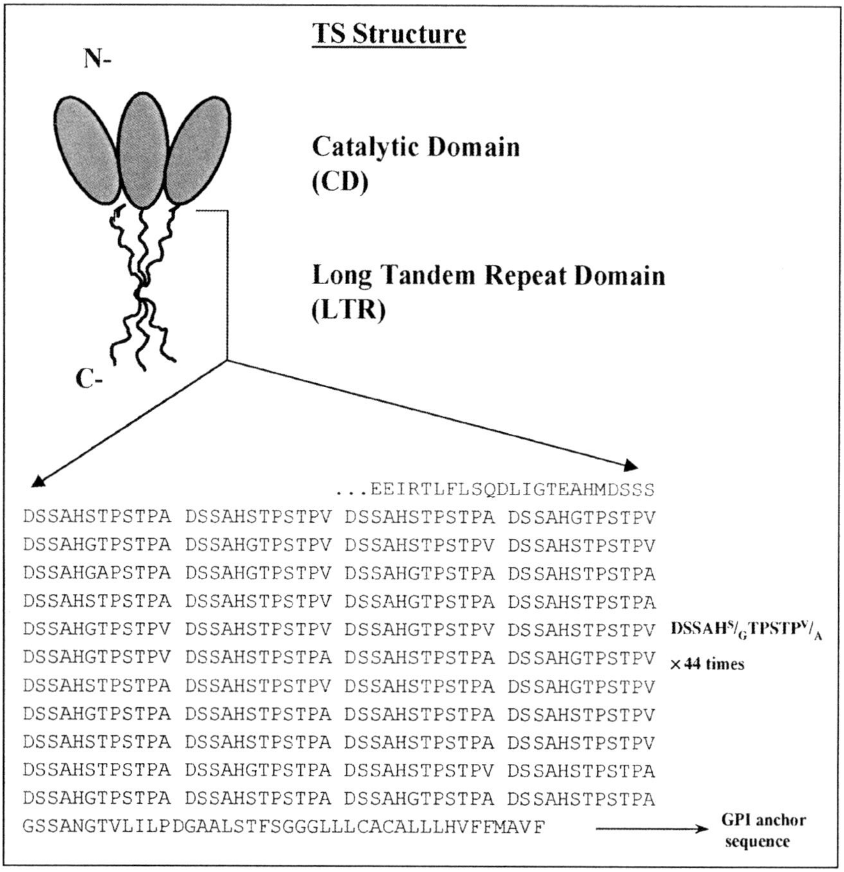

Figure 1. Schematic illustration of TS structure. TS is believed to undergo trimerization under physiological conditions. Each TS molecule contains the N-terminal catalytic domain (CD) and the C-terminal long tandem repeat domain (LTR). Note the LTR is composed of 44 repeats of the highly conserved 12 amino acid sequence (DSSAH^S/$_G$TPSTP^V/$_A$).

TS Is an Activator of B Cell Proliferation and Immunoglobulin Secretion

Authentic cytokines may stimulate not only secretion of other cytokines but also activation and proliferation of immune cells. For example, IL-6, a cytokine whose secretion can be induced by TS in naïve cells, is a potent stimulator of B cells.[64]

Like mitogenic cytokines, TS can induce mitogenesis of lymphocytes. Interestingly, such direct TS effect is specifically on B cells but not on T cells (Fig. 2). TS-induced B cell mitogenesis is independent of T cells and is also mediated by the noncatalytic C-terminal LTR domain of the enzyme (W. Gao, H. Wortis and M. Pereira, manuscript in preparation).

The mechanism of TS-induced B cell mitogenesis is unknown. However, it has been demonstrated that TS activation of B cells requires Bruton's tyrosine kinase (Btk) (Fig. 3), a protein critical to B cell functions.[65-68]

Because secretion of immunoglobulin (Ig) generally follows B cell stimulation by mitogens, one would expect TS-induced B cell proliferation to be accompanied by T cell-independent secretion of Ig nonreactive with TS. Indeed, TS does induce nonspecific Ig secretion in vitro and in vivo (W. Gao, H. Wortis and M. Pereira, manuscript in preparation).

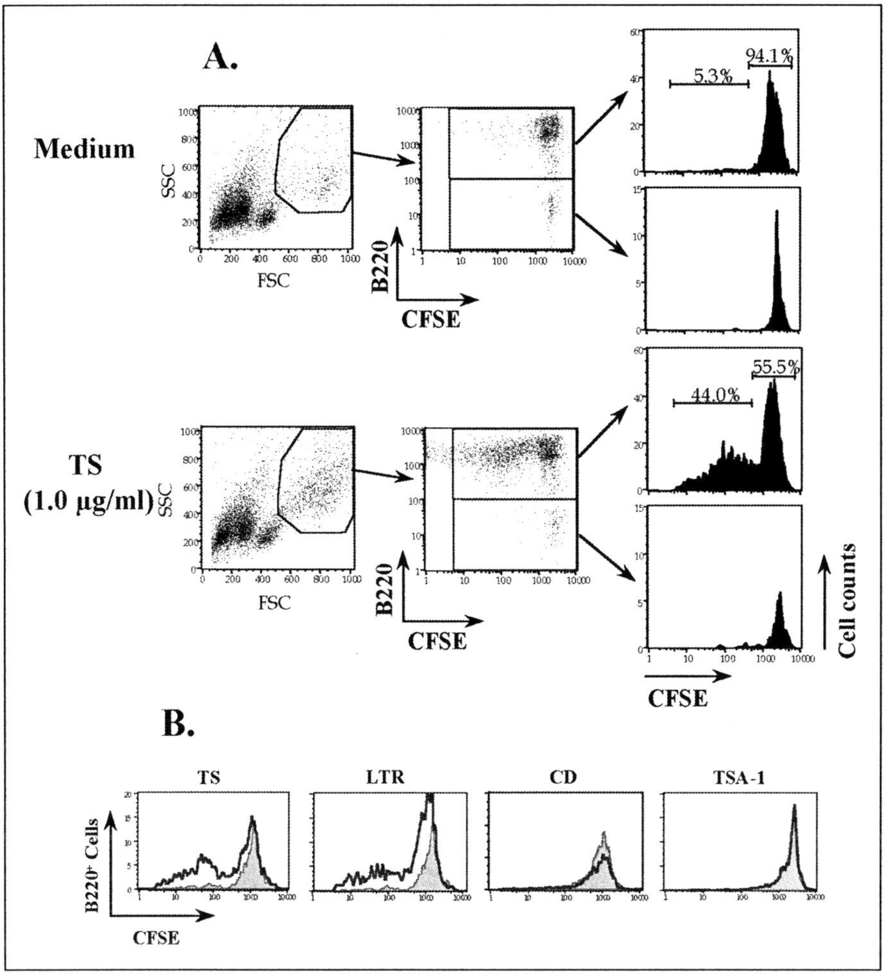

Figure 2. (A). TS stimulated proliferation of B cells (B220$^+$), but not T cells (B220$^-$). Splenocytes from normal C3H/HeJ mice were labeled with carboxyfluorescein diacetate succinimidyl ester (CFSE). The green fluorescent intensity of CFSE-labeled cells reduces by half after each cell division. Three days after TS treatment, there appeared an activated splenocyte population, as shown by FSC and SSC. The activated cells were mainly B220$^+$ B cells. Of those, more than 40% divided multiple times in response to TS, as indicated by the reduced CFSE intensity. (B). TS and LTR, but not CD stimulated B cell proliferation. *T. cruzi* surface antigen-1 (TSA-1) was used as a control. Activated lymphocytes were gated 72 hours after treatment as in (A). CFSE histograms were plotted for activated B220$^+$ cells. Medium control: shaded area; reagent-treated: bold line. Representative data of three different experiments are shown.

TS Potentiates T Cell Activation through Antigen Presenting Cells

TS can potentiate T cell activation triggered by antigen-specific and nonspecific stimuli, though it cannot directly activate T cells.[69] For instance, TS is able to potentiate ConA-induced T cell proliferation and secretion of many cytokines (IL-5, IL-6, IFN-γ and GM-CSF). TS can also strongly enhance antigen-specific T cell activation during immunization. The immuno-potentiating effects of TS on T cells are through its stimulatory activities on the antigen-presenting cells (APC or B cells and macrophages).[69]

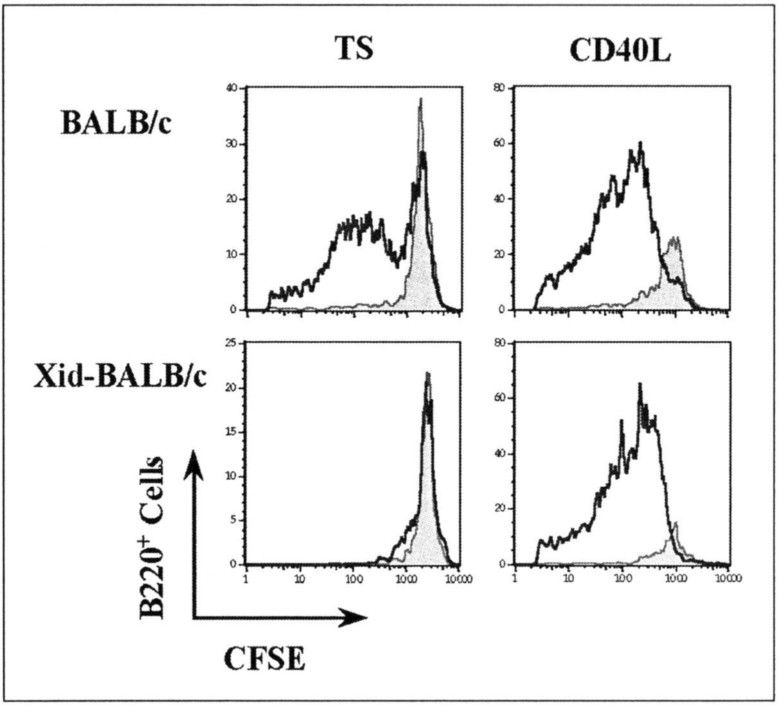

Figure 3. TS stimulation of B cells requires a functional Bruton's tyrosine kinase (Btk). Splenocytes from BALB/c and xid-BALB/c were labeled with CFSE and treated with TS (2.0 μg/ml) or CD40L for 72 hours. Culture supernatant containing the recombinant mouse CD40L-CD8 fusion protein was used as a source of CD40L, which was further crosslinked by anti-CD8. Engagement of CD40 on the B cell surface by CD40L causes equivalent proliferation of wild-type and Btk-defective xid B cells after three days. TS, on the contrary, induced proliferation of only wild-type B cells, but not xid B cells. Activated lymphocytes were gated and CFSE histograms were plotted for B220+ cells. Medium control: shaded area; TS- or CD40L-treated: bold line. Representative data of three different experiments were shown.

Furthermore, TS-mediated potentiation requires IL-6 and Btk.[69] It has been reported that APC-derived IL-6 acts as a costimulatory cytokine for T cell activation.[70,71] In addition, the APC functions of B cells also require a functional Btk.[72] Btk is only expressed in B cells and macrophages.[73,74] Thus, the findings that TS induces APC to secrete IL-6 and to stimulate Btk-dependent B cell activation dovetail nicely with the hypothesis that TS potentiates T cell response through activating antigen-presenting cells (Fig. 4). Therefore, IL-6 and Btk may play important roles in TS-mediated polyclonal activation of B and T cell activation.

Possible Consequences of TS Actions in T. cruzi *Infection*

Polyclonal Lymphocyte Activation

A characteristic feature of the immunological disorder of acute *T. cruzi* infection is polyclonal lymphocyte activation and hypergammaglobulinemia. The majority of activated lymphocytes and secreted Ig do not significantly react with parasite antigens at this stage.[75-77] Consequently, polyclonal lymphocyte activation and hypergammaglobulinemia are believed to underlie the disruption of specific host immunity against the parasite.[78] *T. cruzi,* like many other infectious agents,[47,79] may use this strategy to evade host specific immune attack.

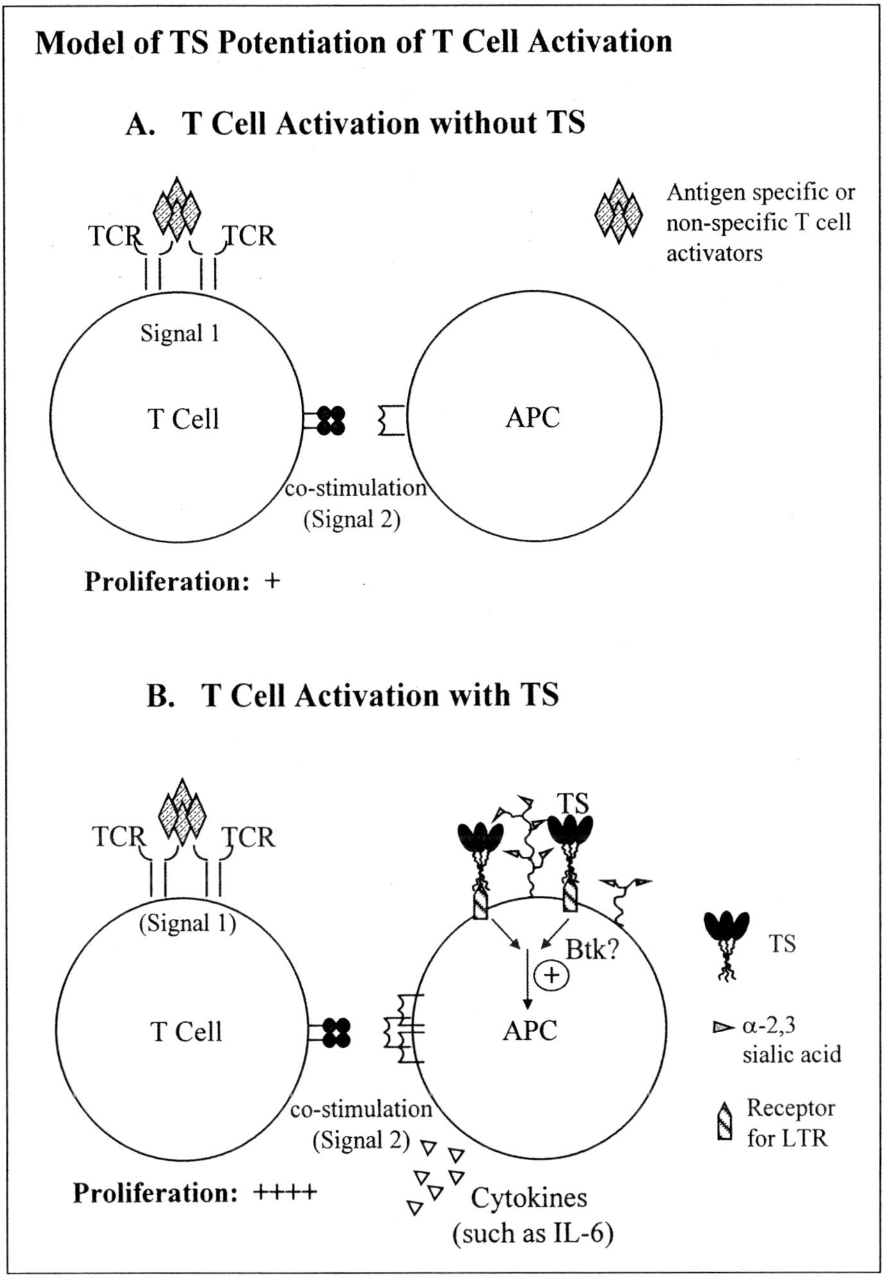

Figure 4. Model for TS potentiation of T cell activation. (A). Optimal T cell activation requires TCR engagement (Signal 1) plus the costimulatory signal (Signal 2). (B). TS, through the long tandem repeat (LTR) domain, binds to an APC surface receptor, whose physiological role is in APC activation and T cell costimulation. TS catalytic domain (CD) mediates the binding of TS to cell surface glycoconjugates, facilitating the LTR to achieve a higher level of receptor cross-linking and a higher level of APC activation. Activated APC then up-regulate more costimulatory molecules and secrete certain cytokines (e.g., IL-6) to further activate T cells.

We postulate TS to be a *T. cruzi* parasitokine that enhances parasite virulence, in part, by driving polyclonal B and T cell activation in acute infection in order to blunt specific anti-parasite responses. Timing is critical for such a strategy to show effectiveness, as the polyclonal lymphocyte activation must be induced before the adaptive immunity takes place. TS is a T cell-independent B cell activator, it rapidly induces polyclonal Ig secretion within the first week of TS exposure in vivo before TS-specific antibodies can be detected (W. Gao, H. Wortis and M. Pereira, manuscript in preparation). Moreover, TS readily induces B cell proliferation and potentiates T cell response independent of the CD40/CD40L pathway,[69] which is crucial for the development of antigen-specific immune responses.[80,81] Thus, TS is capable of inducing polyclonal B and T cell activation within a short period of time before full development of the adaptive immune responses. In the mouse model of Chagas disease, in which animals were sensitized with exogenous TS for infection, TS administrated 1-2 hours prior to parasite inoculation always achieved best virulence enhancement.[60] The in vitro and in vivo activities of TS on lymphocytes might provide a clue for this observation. In a related study, it was shown that administration of IL-2 before or coincident with HIV-gp120 DNA vaccination suppressed gp120-specific responses, whereas administration of IL-2 after vaccination enhanced these responses.[82] Therefore, it may very well be that activation and outgrowth of antigen nonspecific lymphocyte precursors prior to antigen-specific stimulation hinders the development of specific immunity against pathogens.

Based on these observations, it is conceivable that events that suppress polyclonal activation should in theory correlate with increased host survival and reduced tissue damage in *T. cruzi* infection. Two studies in the literature may support this view. The first relates to *T. cruzi* infection in mice deficient in γδ T cells, which exhibit reduced polyclonal lymphocyte activation, mortality rate, and tissue damage compared to the wild-type animals.[83] The second relates to parasite infection in Btk-defective xid mice. When challenged with *T. cruzi*, these mice are highly resistant to infection and exhibit reduced polyclonal lymphocyte activation accompanied with relatively benign tissue pathology.[84]

Although it is not clear at present what exact role Btk may play in the resistance to *T. cruzi* infection, it is noteworthy that polyclonal activation of B cells interferes with host resistance to certain intracellular parasites. For instance, continual administration of anti-IgM antibody, which causes B cell depletion, enhanced resistance to *Leishmania tropica* and *Leishmania mexicana* in BALB/c mice.[85] On the other hand, administration of IL-7, a B cell hematopoietic factor, markedly increased B cell number and exacerbated *L. major* infection.[86] Similarly, cotransfer of B cells converts T cell-reconstituted *L. major* resistant, C.B-17 *scid* mice into a susceptible phenotype.[87] As a B cell mitogen, TS may increase the virulence of TS-transfected *L. major* in a similar fashion. However, the molecular mechanism underlying these observations needs to be further defined.

Activation-Induced Lymphocyte Apoptosis

TS-induced activation of B and T cells could affect lymphocyte homeostasis, especially through activation-induced cell death (AICD). Extensive T cell apoptosis has been observed in the spleens of mice acutely infected by *T. cruzi*.[88-92] B cell apoptosis has also been shown in acute *T. cruzi* infection.[93] AICD is believed to play a role in lymphocyte unresponsiveness to mitogen and antigen in the acute stage,[94,95] and is an important element in undermining the effective control over parasite infection by the host immune system.[92] Supporting this notion, it has been reported that T cell AICD exacerbates *T. cruzi* replication in macrophages.[90,96]

Because AICD is a direct outcome of extended lymphocyte activation,[97,98] it is not unreasonable to postulate that TS accelerates T and B cell AICD. On the other hand, because TS promotes IL-6 secretion and costimulation, both of which can rescue cells from AICD,[99-102] it is equally possible that TS prevents infection-induced AICD and further perpetuates the pathogenic effects of polyclonal lymphocyte activation. Therefore, it would be interesting to determine whether there is a relation between the immuno-potentiating effects of TS and lymphocyte apoptosis. In particular, it will be of interest to compare lymphocyte apoptosis in IL-6 +/+ and IL-6 -/- mice that are infected by *T. cruzi*. The possibility exits that *T. cruzi* infection causes a higher level of lymphocyte apoptosis in IL-6 -/- mice, which will be in keeping with the increased susceptibility in these animals to *T. cruzi* infection (W. Gao and M. Pereira, manuscript in press).

Cytokine Profiles and Cytokine Shock

Cytokine release is an important effector function accompanying lymphocyte activation. Because TS directly stimulates IL-6 secretion from lymphoid and nonlymphoid cells, and potentiates secretion of multiple cytokines from activated T cells, it could disrupt the host cytokine regulation network in vivo through mechanisms independent of antigenic stimulation. Effective control of the intracellular parasite *T. cruzi* relies on the Th1-mediated cellular immunity.[103,104] In our in vitro system, secretion of both type I (IFN-γ) and type II (IL-5 and IL-6) cytokines was significantly augmented by TS in normal mouse splenocytes stimulated by T cell mitogen.[69] It remains to be determined whether TS favors the skewing of Th1/Th2 balance towards a Th2 type in *T. cruzi*-infected animals. If so, it would be in keeping with TS enhancement of *T. cruzi* virulence.

Uncontrolled increase of cytokine secretion, however, might have pathological consequences. For example, during septic shock triggered by bacterial endotoxin, the precipitated release of significant levels of pro-inflammatory cytokines could lead to host morbidity and mortality.[105] These cytokines include TNF-α, IFN-γ, IL-12 and IL-18,[106] whose elevated levels cause sequential multiple organ failure and host lethality. Viral or bacterial superantigens can induce cytokine shock.[48] A similar phenomenon has been reported for parasitic infection in certain strains of cytokine-deficient mice.[107-110] For instance, mice deficient in the Th2 cytokine IL-10 were expected to be more efficient in controlling the intracellular parasites *T. gondii* and *T. cruzi*, and hence they might have better chances to survive the infections.[111] However, when IL-10 -/- mice were infected with *T. gondii*, they quickly succumbed to a T cell-mediated shock-like reaction characterized by the overproduction of IL-12 and IFN-γ, associated with widespread necrosis in the liver, a reaction not seen in the wild-type animals.[107] Similarly, these mice died more promptly than the wild-type animals upon *T. cruzi* challenge. The levels of IFN-γ, TNF-α and IL-12 in these mice were 1-2 logs higher than those in the wild-type animals during infection.[109,110] Cytokine toxicity was responsible for the morbidity and mortality of IL-10 -/- mice because much fewer *T. gondii* or *T. cruzi* parasites were detected in IL-10 -/- than in the wild-type mice. In addition, neutralizing antibodies to IFN-γ, TNF-α and IL-12 prevented the early death of infected IL-10 -/- mice.[107,109,110] It has also been shown that cytokine shock could be mediated through enhanced costimulatory activities of antigen-presenting cells.[112,113]

Based on the effects of TS on cytokine induction and APC activation, it is possible that TS compromises the physiological fitness of the host by inducing cytokine shock, thereby enhancing parasite virulence.[60] Clearly, further analysis needs to be carried out to characterize the role of TS in modulating the host Th cytokine profile and in inducing cytokine toxicity during *T. cruzi* infection.

Other Parasitokines

T. cruzi *Proline Racemase*

T. cruzi secretes other parasitokines in addition to TS. It has been established that the parasite secretes at least two other B cell mitogens: a 24 kDa protein and a proline racemase.[114,115]

Like TS, the two mitogenic *T. cruzi* proteins are able to directly activate normal B cells from naïve mice (see Chapter 7 , P. Minoprio).

T Lymphocyte Triggering Factor (TLTF) from T. brucei

A well-studied parasitokine is the T lymphocyte triggering factor (TLTF) from *T. brucei*, the protozoan parasite that causes African sleeping sickness. TLTF polyclonally activates $CD8^+$, but not $CD4^+$ T lymphocytes to proliferate and to secrete IFN-γ.[116-118] Even though relative resistance to African trypanosomes is associated with a strong IFN-γ-dependent Th1 response to parasite antigens,[119] *T. brucei* strikingly utilizes IFN-γ as a parasite growth factor.[116-118,120] This may help explain the paradoxical observations that mice with a disrupted IFN-γ gene have reduced parasitemia and a prolonged survival to *T. brucei* infection, while the outcome is reversed in mice that lack the IFN-γ receptor gene.[120]

Mimic of Macrophage Migration Inhibitory Factor from Brugia malayi *(Bm-MIF)*

The microfilariae *B. malayi* secretes a parasitokine (Bm-MIF) that mimics the pro-inflammatory cytokine macrophage migration inhibitory factor (MIF).[121] MIF is a mediator of several diseases including gram-negative septic shock and delayed-type hypersensitivity reactions. One of the immunological functions of MIF is to modulate the host macrophage and T and B cell response.[122] Bm-MIF is a 12.5 kDa peptide exhibiting 42% identity to human and murine MIF and has the capability of modifying the activities of human monocytes/macrophages.[121] The precise mechanism of the parasitokine Bm-MIF in sabotaging host immune responses to promote parasite survival remains to be determined.

Applications of Pathogen-Derived Cytokine Mimetics

Modern medicine relies heavily on the discoveries that facilitate modulation and exploitation of the host immune system. Pathogens striving for their biological niches in the hosts have acquired the capability of such modulation and exploitation during evolution. Thus, pathogen-derived cytokine mimetics are the evolution-tested, effectiveness-proven entities that may be developed into valuable therapeutic reagents.

For instance, viral IL-10 (vIL-10) from EBV shares many of the anti-inflammatory properties of mouse and human IL-10, but lacks their immuno-stimulatory properties and may therefore offer superior immunosuppression.[123,124] Different groups have reported that vIL-10 delivered by adenovirus-mediated gene transfer delayed the onset and reduced the incidence and severity of collagen-induced arthritis (CIA) in animal models.[123-126] The strong effect of vIL-10 is systemic, as it suppresses the development of CIA in both injected and uninjected contralateral joints.[125,126] In addition, lipid-mediated gene transfer of vIL-10 has also been shown to prolong the survival of vascularized cardiac allografts by inhibiting donor-specific cellular and humoral immune responses.[127] Thus, vIL-10 could be a valuable immunosuppressant for treating human autoimmune and transplantation-related diseases.

On the other hand, pathogen-derived cytokine mimetics with immuno-potentiating properties may be used as adjuvant for immunization or treatment of certain immune deficient diseases. For example, it has been widely reported that cholera toxin (CT), *E. coli* heat-labile enterotoxin (LT), *Pertussis* toxin (PT) and other bacterial toxins are strong adjuvants that boost both humoral and cellular responses during mucosal immunization.[128-133] The adjuvant effect of these bacteriokines is associated with induced production of regulatory cytokines and expression of the costimulatory molecules B7-1 and B7-2.[130,133] Similar to these effects, the *T. cruzi* TS, through its LTR domain, boosts both cellular and humoral responses. Therefore, the idea that parasitokines can be exploited for clinical purposes is highly intriguing and clearly warrants future investigation.

References

1. Wurzner R. Evasion of pathogens by avoiding recognition or eradication by complement, in part via molecular mimicry. Mol Immunol 1999; 36:249-260.
2. Abu-Shakra M, Buskila D, Shoenfeld Y. Molecular mimicry between host and pathogen: examples from parasites and implication. Immunol Lett 1999; 67:147-152.
3. Newbold CI. Antigenic variation in Plasmodium falciparum: mechanisms and consequences. Curr Opin Microbiol 1999; 2:420-425.
4. Pays E, Vanhamme L, Berberof M. Genetic controls for the expression of surface antigens in African trypanosomes. Annu Rev Microbiol 1994; 48:25-52.
5. Galan JE, Collmer A. Type III secretion machines: bacterial devices for protein delivery into host cells. Science 1999; 284:1322-1328.
6. Anderson DM, Schneewind O. Type III machines of Gram-negative pathogens: injecting virulence factors into host cells and more. Curr Opin Microbiol 1999; 2:18-24.
7. Cornelis GR, Wolf-Watz H. The Yersinia Yop virulon: a bacterial system for subverting eukaryotic cells. Mol Microbiol 1997; 23:861-867.
8. DeVinney I, Steele-Mortimer I, Finlay BB. Phosphatases and kinases delivered to the host cell by bacterial pathogens. Trends Microbiol 2000;8:29-33.
9. Orth K, Palmer LE, Bao ZQ et al. Inhibition of the mitogen-activated protein kinase kinase superfamily by a Yersinia effector. Science 1999; 285:1920-1923.
10. Aepfelbacher M, Zumbihl R, Ruckdeschel K et al. The tranquilizing injection of Yersinia proteins: a pathogen's strategy to resist host defense. Biol Chem 1999; 380:795-802.
11. Ruckdeschel K, Roggenkamp A, Lafont V et al. Interaction of Yersinia enterocolitica with macrophages leads to macrophage cell death through apoptosis. Infect Immun 1997; 65:4813-4821.
12. Mills SD, Boland A, Sory MP et al. Yersinia enterocolitica induces apoptosis in macrophages by a process requiring functional type III secretion and translocation mechanisms and involving YopP, presumably acting as an effector protein. Proc Natl Acad Sci USA 1997; 94:12638-12643.
13. Ruckdeschel K, Harb S, Roggenkamp A et al. Yersinia enterocolitica impairs activation of transcription factor NF-kappaB: involvement in the induction of programmed cell death and in the suppression of the macrophage tumor necrosis factor alpha production. J Exp Med 1998; 187:1069-1079.
14. Suzuki Y, Orellana MA, Schreiber RD et al. Interferon-gamma: the major mediator of resistance against Toxoplasma gondii. Science 1988; 240:516-518.
15. Gazzinelli RT, Wysocka M, Hayashi S et al. Parasite-induced IL-12 stimulates early IFN-gamma synthesis and resistance during acute infection with Toxoplasma gondii. J Immunol 1994; 153:2533-2543.
16. Denkers EY, Yap G, Scharton-Kersten T et al. Perforin-mediated cytolysis plays a limited role in host resistance to Toxoplasma gondii. J Immunol 1997; 159:1903-1908.
17. Suzuki Y, Remington JS. The effect of anti-IFN-gamma antibody on the protective effect of Lyt-2+ immune T cells against toxoplasmosis in mice. J Immunol 1990; 144:1954-1956.
18. Feghali CA, Wright TM. Cytokines in acute and chronic inflammation. Front Biosci 1997; 2:d12-26.
19. Gauldie J, Northemann W, Fey GH. IL-6 functions as an exocrine hormone in inflammation. Hepatocytes undergoing acute phase responses require exogenous IL-6. J Immunol 1990; 144:3804-3808.
20. Kozak W, Zheng H, Conn CA et al. Thermal and behavioral effects of lipopolysaccharide and influenza in interleukin-1 beta-deficient mice. Am J Physiol 1995; 269:R969-977.
21. Castanos-Velez E, Maerlan S, Osorio LM et al. Trypanosoma cruzi infection in tumor necrosis factor receptor p55-deficient mice. Infect Immun 1998; 66:2960-2968.
22. Rothe J, Lesslauer W, Lotscher H et al. Mice lacking the tumour necrosis factor receptor 1 are resistant to TNF-mediated toxicity but highly susceptible to infection by Listeria monocytogenes. Nature 1993; 364:798-802.
23. Kopf M, Baumann H, Freer G et al. Impaired immune and acute-phase responses in interleukin-6-deficient mice. Nature 1994; 368:339-342.
24. Kotwal GJ. Virokines: mediators of virus-host interaction and future immunomodulators in medicine. Arch Immunol Ther Exp 1999; 47:135-138.

25. Wilson M, Seymour R, Henderson B. Bacterial perturbation of cytokine networks. Infect Immun 1998; 66:2401-2409.
26. McFadden G, Graham K, Ellison K et al. Interruption of cytokine networks by poxviruses: lessons from myxoma virus. J Leukoc Biol 1995; 57:731-738.
27. Moore PS, Boshoff C, Weiss RA et al. Molecular mimicry of human cytokine and cytokine response pathway genes by KSHV. Science 1996; 274:1739-1744.
28. Neipel F, Albrecht JC, Ensser A et al. Human herpesvirus 8 encodes a homolog of interleukin-6. J Virol 1997; 71:839-842.
29. Moore KW, Vieira P, Fiorentino DF et al. Homology of cytokine synthesis inhibitory factor (IL-10) to the Epstein-Barr virus gene BCRFI. Science 1990; 248:1230-1234.
30. Fleming SB, McCaughan CA, Andrews AE et al. A homolog of interleukin-10 is encoded by the poxvirus orf virus. J Virol 1997; 71:4857-4861.
31. Rode HJ, Janssen W, Rosen-Wolff A et al. The genome of equine herpesvirus type 2 harbors an interleukin 10 (IL10)-like gene. Virus Genes 1993; 7:111-116.
32. Yao Z, Fanslow WC, Seldin MF et al. Herpesvirus Saimiri encodes a new cytokine, IL-17, which binds to a novel cytokine receptor. Immunity 1995; 3:811-821.
33. Upton C, Macen JL, Schreiber M et al. Myxoma virus expresses a secreted protein with homology to the tumor necrosis factor receptor gene family that contributes to viral virulence. Virology 1991; 184:370-382.
34. Smith CA, Davis T, Wignall JM et al. T2 open reading frame from the Shope fibroma virus encodes a soluble form of the TNF receptor. Biochem Biophys Res Commun 1991; 176:335-342.
35. Hu FQ, Smith CA, Pickup DJ. Cowpox virus contains two copies of an early gene encoding a soluble secreted form of the type II TNF receptor. Virology 1994; 204:343-356.
36. Smith VP, Alcami A. Expression of secreted cytokine and chemokine inhibitors by ectromelia virus. J Virol 2000; 74:8460-8471.
37. Saraiva M, Alcami A. CrmE, a novel soluble tumor necrosis factor receptor encoded by poxviruses. J Virol 2001; 75:226-233.
38. Upton C, Mossman K, McFadden G. Encoding of a homolog of the IFN-gamma receptor by myxoma virus. Science 1992; 258:1369-1372.
39. Alcami A, Smith GL. Vaccinia, cowpox, and camelpox viruses encode soluble gamma interferon receptors with novel broad species specificity. J Virol 1995; 69:4633-4639.
40. Colamonici OR, Domanski P, Sweitzer SM et al. Vaccinia virus B18R gene encodes a type I interferon-binding protein that blocks interferon alpha transmembrane signaling. J Biol Chem 1995; 270:15974-15978.
41. Alcami A, Smith GL. A soluble receptor for interleukin-1 beta encoded by vaccinia virus: a novel mechanism of virus modulation of the host response to infection. Cell 1992; 71:153-167.
42. Xiang Y, Moss B. IL-18 binding and inhibition of interferon gamma induction by human poxvirus-encoded proteins. Proc Natl Acad Sci USA 1999; 96:11537-11542.
43. Smith VP, Bryant NA, Alcami A. Ectromelia, vaccinia and cowpox viruses encode secreted interleukin-18-binding proteins. J Gen Virol 2000; 81(Pt 5):1223-1230.
44. Aoki Y, Jaffe ES, Chang Y et al. Angiogenesis and hematopoiesis induced by Kaposi's sarcoma-associated herpesvirus-encoded interleukin-6. Blood 1999; 93:4034-4043.
45. Mossman K, Nation P, Macen J et al. Myxoma virus M-T7, a secreted homolog of the interferon-gamma receptor, is a critical virulence factor for the development of myxomatosis in European rabbits. Virology 1996; 215:17-30.
46. Henderson B, Poole S, Wilson M. Microbial/host interactions in health and disease: who controls the cytokine network? Immunopharmacology 1996; 35:1-21.
47. Fleischer B. Superantigens produced by infectious pathogens: molecular mechanism of action and biological significance. Int J Clin Lab Res 1994; 24:193-197.
48. Miethke T, Wahl C, Regele D et al. Superantigen mediated shock: a cytokine release syndrome. Immunobiology 1993; 189:270-284.
49. Baschieri S, Lees RK, Lussow AR et al. Clonal anergy to staphylococcal enterotoxin B in vivo: selective effects on T cell subsets and lymphokines. Eur J Immunol 1993; 23:2661-2666.

50. Perkins DL, Wang Y, Ho SS et al. Superantigen-induced peripheral tolerance inhibits T cell responses to immunogenic peptides in TCR (beta-chain) transgenic mice. J Immunol 1993; 150:4284-4291.
51. Ettinger R, Panka DJ, Wang JK et al. Fas ligand-mediated cytotoxicity is directly responsible for apoptosis of normal CD4+ T cells responding to a bacterial superantigen. J Immunol 1995; 154:4302-4308.
52. Pereira ME. A developmentally regulated neuraminidase activity in Trypanosoma cruzi. Science 1983; 219:1444-1446.
53. Prioli RP, Mejia JS, Aji T et al. Trypanosoma cruzi: localization of neuraminidase on the surface of trypomastigotes. Trop Med Parasitol 1991; 42:146-150.
54. Scudder P, Doom JP, Chuenkova M et al. Enzymatic characterization of beta-D-galactoside alpha 2,3-trans-sialidase from Trypanosoma cruzi. J Biol Chem 1993; 268:9886-9891.
55. Schenkman S, Eichinger D, Pereira ME et al. Structural and functional properties of Trypanosoma trans-sialidase. Annu Rev Microbiol 1994; 48:499-523.
56. Pereira ME, Mejia JS, Ortega-Barria E et al. The Trypanosoma cruzi neuraminidase contains sequences similar to bacterial neuraminidases, YWTD repeats of the low density lipoprotein receptor, and type III modules of fibronectin. J Exp Med 1991; 174:179-191.
57. Ming M, Chuenkova M, Ortega-Barria E et al. Mediation of Trypanosoma cruzi invasion by sialic acid on the host cell and trans-sialidase on the trypanosome. Mol Biochem Parasitol 1993; 59:243-252.
58. Schenkman RP, Vandekerckhove F, Schenkman S. Mammalian cell sialic acid enhances invasion by Trypanosoma cruzi. Infect Immun 1993; 61:898-902.
59. Pereira ME, Zhang K, Gong Y et al. Invasive phenotype of Trypanosoma cruzi restricted to a population expressing trans-sialidase. Infect Immun 1996; 64:3884-3892.
60. Chuenkova M, Pereira ME. Trypanosoma cruzi trans-sialidase: enhancement of virulence in a murine model of Chagas disease. J Exp Med 1995; 181:1693-1703.
61. Belen Carrillo M, Gao W, Herrera M et al. Heterologous expression of Trypanosoma cruzi trans-sialidase in Leishmania major enhances virulence. Infect Immun 2000; 68:2728-2734.
62. Saavedra E, Herrera M, Gao W et al. The Trypanosoma cruzi trans-sialidase, through its COOH-terminal tandem repeat, upregulates interleukin 6 secretion in normal human intestinal microvascular endothelial cells and peripheral blood mononuclear cells. J Exp Med 1999; 190:1825-1836.
63. Dinarello CA. Interleukin-1 and its biologically related cytokines. Adv Immunol 1989; 44:153-205.
64. Hirano T, Akira S, Taga T et al. Biological and clinical aspects of interleukin 6. Immunol Today 1990; 11:443-449.
65. Bajpai UD, Zhang K, Teutsch M et al. Bruton's tyrosine kinase links the B cell receptor to nuclear factor kappaB activation. J Exp Med 2000; 191:1735-1744.
66. Petro JB, Rahman SM, Ballard DW et al. Bruton's tyrosine kinase is required for activation of IkappaB kinase and nuclear factor kappaB in response to B cell receptor engagement. J Exp Med 2000; 191:1745-1754.
67. Santos-Argumedo L, Lund FE, Heath AW et al. CD38 unresponsiveness of xid B cells implicates Bruton's tyrosine kinase (btk) as a regular of CD38 induced signal transduction. Int Immunol 1995; 7:163-170.
68. Koike M, Kikuchi Y, Tominaga A et al. Defective IL-5-receptor-mediated signaling in B cells of X-linked immunodeficient mice. Int Immunol 1995; 7:21-30.
69. Gao W, Pereira MA. Trypanosoma cruzi trans-Sialidase Potentiates T Cell Activation through Antigen-Presenting Cells: Role of IL-6 and Bruton's Tyrosine Kinase. Eur J Immunol 2001; 31:1503-1512.
70. Tosato G, Pike SE. Interferon-beta 2/interleukin 6 is a co-stimulant for human T lymphocytes. J Immunol 1988; 141:1556-1562.
71. Mincheff MS, Meryman HT. Costimulatory signals necessary for induction of T cell proliferation. Transplantation 1990; 49:768-772.
72. Goldstein MD, Debenedette MA, Hollenbaugh D et al. Induction of costimulatory molecules B7-1 and B7-2 in murine B cells. the CBA/N mouse reveals a role for Bruton's tyrosine kinase in CD40-mediated B7 induction. Mol Immunol 1996; 33:541-552.

73. Genevier HC, Hinshelwood S, Gaspar HB et al. Expression of Bruton's tyrosine kinase protein within the B cell lineage. Eur J Immunol 1994; 24:3100-3105.
74. Smith CI, Baskin B, Humire-Greiff P et al. Expression of Bruton's agammaglobulinemia tyrosine kinase gene, BTK, is selectively down-regulated in T lymphocytes and plasma cells. J Immunol 1994; 152:557-565.
75. Minoprio PM, Eisen H, Forni L et al. Polyclonal lymphocyte responses to murine Trypanosoma cruzi infection. I. Quantitation of both T- and B-cell responses. Scand J Immunol 1986; 24:661-668.
76. Minoprio P, Burlen O, Pereira P et al. Most B cells in acute Trypanosoma cruzi infection lack parasite specificity. Scand J Immunol 1988; 28:553-561.
77. Minoprio P, Itohara S, Heusser C et al. Immunobiology of murine T. cruzi infection: the predominance of parasite-nonspecific responses and the activation of TCRI T cells. Immunol Rev 1989; 112:183-207.
78. DosReis GA. Cell-mediated Immunity in Experimental Trypanosoma cruzi Infection. Parasitol Today 1997; 13:335-342.
79. Galvao-Castro B, Sa Ferreira JA, Marzochi KF et al. Polyclonal B cell activation, circulating immune complexes and autoimmunity in human american visceral leishmaniasis. Clin Exp Immunol 1984; 56:58-66.
80. Hollenbaugh D, Ochs HD, Noelle RJ et al. The role of CD40 and its ligand in the regulation of the immune response. Immunol Rev 1994; 138:23-37.
81. Marshall LS, Aruffo A, Ledbetter JA et al. The molecular basis for T cell help in humoral immunity: CD40 and its ligand, gp39. J Clin Immunol 1993; 13:165-174.
82. Barouch DH, Santra S, Steenbeke TD et al. Augmentation and suppression of immune responses to an HIV-1 DNA vaccine by plasmid cytokine/Ig administration. J Immunol 1998; 161:1875-1882.
83. Santos Lima EC, Minoprio P. Chagas disease is attenuated in mice lacking gamma delta T cells. Infect Immun 1996; 64:215-221.
84. Minoprio P, Coutinho A, Spinella S et al. Xid immunodeficiency imparts increased parasite clearance and resistance to pathology in experimental Chagas disease. Int Immunol 1991; 3:427-433.
85. Sacks DL, Scott PA, Asofsky R et al. Cutaneous leishmaniasis in anti-IgM-treated mice: enhanced resistance due to functional depletion of a B cell-dependent T cell involved in the suppressor pathway. J Immunol 1984; 132:2072-2077.
86. Hoerauf A, Solbach W, Rollinghoff M et al. Effect of IL-7 treatment on Leishmania major-infected BALB.Xid mice: enhanced lymphopoiesis with sustained lack of B1 cells and clinical aggravation of disease. Int Immunol 1995; 7:1879-1884.
87. Hoerauf A, Rollinghoff M, Solbach W. Co-transfer of B cells converts resistance into susceptibility in T cell-reconstituted, Leishmania major-resistant C.B-17 scid mice by a non-cognate mechanism. Int Immunol 1996; 8:1569-1575.
88. DosReis GA, Fonseca MEF, Lopes MF. Programmed T-cell Death in Experimental Chagas Disease. Parasitol Today 1995; 11:390-394.
89. Lopes MF, da Veiga VF, Santos AR et al. Activation-induced CD4+ T cell death by apoptosis in experimental Chagas disease. J Immunol 1995; 154:744-752.
90. Nunes MP, Andrade RM, Lopes MF et al. Activation-induced T cell death exacerbates Trypanosoma cruzi replication in macrophages cocultured with CD4+ T lymphocytes from infected hosts. J Immunol 1998; 160:1313-1319.
91. Lopes MF, DosReis GA. Trypanosoma cruzi-induced immunosuppression: selective triggering of CD4+ T-cell death by the T-cell receptor-CD3 pathway and not by the CD69 or Ly-6 activation pathway. Infect Immun 1996; 64:1559-1564.
92. Barcinski MA, DosReis GA. Apoptosis in parasites and parasite-induced apoptosis in the host immune system: a new approach to parasitic diseases. Braz J Med Biol Res 1999; 32:395-401.
93. Zuniga E, Motran C, Montes CL et al. Trypanosoma cruzi-induced immunosuppression: B cells undergo spontaneous apoptosis and lipopolysaccharide (LPS) arrests their proliferation during acute infection. Clin Exp Immunol 2000; 119:507-515.
94. Kierszenbaum F, Hayes MM. Evaluation of lymphocyte responsiveness to polyclonal activators during acute and chronic experimental Trypanosoma cruzi infection. Am J Trop Med Hyg 1980; 29:708-710.

95. Lopes MF, DosReis GA. Apoptosis as a cause of T-cell unresponsiveness in experimental Chagas disease. Braz J Med Biol Res 1995; 28:913-918.
96. Freire-de-Lima CG, Nascimento DO, Soares MB et al. Uptake of apoptotic cells drives the growth of a pathogenic trypanosome in macrophages. Nature 2000; 403:199-203.
97. Kabelitz D, Janssen O. Antigen-induced death of T-Lymphocytes. Front Biosci 1997; 2:d61-77.
98. Wesselborg S, Janssen O, Kabelitz D. Induction of activation-driven death (apoptosis) in activated but not resting peripheral blood T cells. J Immunol 1993; 150:4338-4345.
99. Kishimoto H, Sprent J. Strong TCR ligation without costimulation causes rapid onset of Fas-dependent apoptosis of naive murine CD4+ T cells. J Immunol 1999; 163:1817-1826.
100. Sepulveda H, Cerwenka A, Morgan T et al. CD28, IL-2-independent costimulatory pathways for CD8 T lymphocyte activation. J Immunol 1999; 163:1133-1142.
101. Ayroldi E, Zollo O, Cannarile L et al. Interleukin-6 (IL-6) prevents activation-induced cell death: IL-2- independent inhibition of Fas/FasL expression and cell death. Blood 1998; 92:4212-4219.
102. Daniel PT, Scholz C, Westermann J et al. Dendritic cells prevent CD95 mediated T lymphocyte death through costimulatory signals. Adv Exp Med Biol 1998; 451:173-177.
103. Hoft DF, Schnapp AR, Eickhoff CS et al. Involvement of CD4(+) Th1 cells in systemic immunity protective against primary and secondary challenges with Trypanosoma cruzi. Infect Immun 2000; 68:197-204.
104. Tarleton RL, Grusby MJ, Zhang L. Increased susceptibility of Stat4-deficient and enhanced resistance in Stat6-deficient mice to infection with Trypanosoma cruzi. J Immunol 2000; 165:1520-1525.
105. Karima R, Matsumoto S, Higashi H et al. The molecular pathogenesis of endotoxic shock and organ failure. Mol Med Today 1999; 5:123-132.
106. Hochholzer P, Lipford GB, Wagner H et al. Role of interleukin-18 (IL-18) during lethal shock: decreased lipopolysaccharide sensitivity but normal superantigen reaction in IL-18-deficient mice. Infect Immun 2000; 68:3502-3508.
107. Gazzinelli RT, Wysocka M, Hieny S et al. In the absence of endogenous IL-10, mice acutely infected with Toxoplasma gondii succumb to a lethal immune response dependent on CD4+ T cells and accompanied by overproduction of IL-12, IFN-gamma and TNF-alpha. J Immunol 1996; 157:798-805.
108. Roberts CW, Ferguson DJ, Jebbari H et al. Different roles for interleukin-4 during the course of Toxoplasma gondii infection. Infect Immun 1996; 64:897-904.
109. Hunter CA, Ellis-Neyes LA, Slifer T et al. IL-10 is required to prevent immune hyperactivity during infection with Trypanosoma cruzi. J Immunol 1997; 158:3311-3316.
110. Holscher C, Mohrs M, Dai WJ et al. Tumor necrosis factor alpha-mediated toxic shock in Trypanosoma cruzi-infected interleukin 10-deficient mice. Infect Immun 2000; 68:4075-4083.
111. Reed SG, Brownell CE, Russo DM et al. IL-10 mediates susceptibility to Trypanosoma cruzi infection. J Immunol 1994; 153:3135-3140.
112. Villegas EN, Wille U, Craig L et al. Blockade of costimulation prevents infection-induced immunopathology in interleukin-10-deficient mice. Infect Immun 2000; 68:2837-2844.
113. Reis e Sousa C, Yap G, Schulz O et al. Paralysis of dendritic cell IL-12 production by microbial products prevents infection-induced immunopathology. Immunity 1999; 11:637-647.
114. Da Silva AC, Espinoza AG, Taibi A et al. A 24,000 MW Trypanosoma cruzi antigen is a B-cell activator. Immunology 1998; 94:189-196.
115. Reina-San-Martin B, Degrave W, Rougeot C et al. A B-cell mitogen from a pathogenic trypanosome is a eukaryotic proline racemase. Nat Med 2000; 6:890-897.
116. Olsson T, Bakhiet M, Edlund C et al. Bidirectional activating signals between Trypanosoma brucei and CD8+ T cells: a trypanosome-released factor triggers interferon-gamma production that stimulates parasite growth. Eur J Immunol 1991; 21:2447-2454.
117. Olsson T, Bakhiet M, Hojeberg B et al. CD8 is critically involved in lymphocyte activation by a T. brucei brucei-released molecule. Cell 1993; 72:715-727.
118. Vaidya T, Bakhiet M, Hill KL et al. The gene for a T lymphocyte triggering factor from African trypanosomes. J Exp Med 1997; 186:433-438.
119. Hertz CJ, Filutowicz H, Mansfield JM. Resistance to the African trypanosomes is IFN-gamma dependent. J Immunol 1998; 161:6775-6783.

120. Bakhiet M, Olsson T, Mhlanga J et al. Human and rodent interferon-gamma as a growth factor for Trypanosoma brucei. Eur J Immunol 1996; 26:1359-1364.
121. Pastrana DV, Raghavan N, FitzGerald P et al. Filarial nematode parasites secrete a homologue of the human cytokine macrophage migration inhibitory factor. Infect Immun 1998; 66:5955-5963.
122. Bernhagen J, Calandra T, Bucala R. Regulation of the immune response by macrophage migration inhibitory factor: biological and structural features. J Mol Med 1998; 76:151-161.
123. Ma Y, Thornton S, Duwel LE et al. Inhibition of collagen-induced arthritis in mice by viral IL-10 gene transfer. J Immunol 1998; 161:1516-1524.
124. Apparailly F, Verwaerde C, Jacquet C et al. Adenovirus-mediated transfer of viral IL-10 gene inhibits murine collagen-induced arthritis. J Immunol 1998; 160:5213-5220.
125. Whalen JD, Lechman EL, Carlos CA et al. Adenoviral transfer of the viral IL-10 gene periarticularly to mouse paws suppresses development of collagen-induced arthritis in both injected and uninjected paws. J Immunol 1999; 162:3625-3632.
126. Lechman ER, Jaffurs D, Ghivizzani SC et al. Direct adenoviral gene transfer of viral IL-10 to rabbit knees with experimental arthritis ameliorates disease in both injected and contralateral control knees. J Immunol 1999; 163:2202-2208.
127. DeBruyne LA, Li K, Chan SY et al. Lipid-mediated gene transfer of viral IL-10 prolongs vascularized cardiac allograft survival by inhibiting donor-specific cellular and humoral immune responses. Gene Ther 1998; 5:1079-1087.
128. Lebens M, Holmgren J. Mucosal vaccines based on the use of cholera toxin B subunit as immunogen and antigen carrier. Dev Biol Stand 1994; 82:215-227.
129. Freytag LC, Clements JD. Bacterial toxins as mucosal adjuvants. Curr Top Microbiol Immunol 1999; 236:215-236.
130. Yamamoto M, McGhee JR, Hagiwara Y et al. Genetically manipulated bacterial toxin as a new generation mucosal adjuvant. Scand J Immunol 2001; 53:211-217.
131. Simmons CP, Hussell T, Sparer T et al. Mucosal delivery of a respiratory syncytial virus CTL peptide with enterotoxin-based adjuvants elicits protective, immunopathogenic, and immunoregulatory antiviral CD8+ T cell responses. J Immunol 2001; 166:1106-1113.
132. Simmons CP, Mastroeni P, Fowler R et al. MHC class I-restricted cytotoxic lymphocyte responses induced by enterotoxin-based mucosal adjuvants. J Immunol 1999; 163:6502-6510.
133. Ryan M, McCarthy L, Rappuoli R et al. Pertussis toxin potentiates Th1 and Th2 responses to co-injected antigen: adjuvant action is associated with enhanced regulatory cytokine production and expression of the co-stimulatory molecules B7-1, B7-2 and CD28. Int Immunol 1998; 10:651-662.

Index

I

K

L

M

N

P

R

S

T

V

Z